Synthesis Lectures on Operations Research and Applications

Series Editor

James Cochran, Ames, USA

This series focuses on the use of advanced analytics in both industry and scientific research to advance the quality of decisions and processes. Written by international experts, modern applications and methodologies are utilized to help researchers and students alike to improve their use of analytics. Classical and cutting-edge topics are presented and explored with a focus on utilization and application across a range in practical situations.

Özgün Elçi · John Hooker · Peter Zhang

The Structure of Fair Solutions

Achieving Fairness in an Optimization Model

Springer

Özgün Elçi
Amazon
Seattle, WA, USA

John Hooker
Tepper School of Business
Carnegie Mellon University
Pittsburgh, PA, USA

Peter Zhang
Carnegie Mellon University
Pittsburgh, PA, USA

ISSN 2770-6303 ISSN 2770-6311 (electronic)
Synthesis Lectures on Operations Research and Applications
ISBN 978-3-031-82189-9 ISBN 978-3-031-82190-5 (eBook)
https://doi.org/10.1007/978-3-031-82190-5

This Springer imprint is published by the registered company Springer Nature Switzerland AG
The registered company address is: Gewerbestrasse 11, 6330 Cham, Switzerland

Preface

Fairness is often an important consideration in the optimization models that lie behind business and public policy decisions. Yet how can fairness be defined mathematically? A number of fairness models have been proposed, but one must somehow select the right model for a given application. In this book, we address this challenge by deducing general characteristics of optimal solutions that result from various fairness models. One can then judge which type of solution is acceptable in a given context. We focus on fairness criteria that combine efficiency with fairness, since typically both are important in practice. Most of the results are new, and some are quite unobvious and even surprising.

Our aim has been to provide a concise but mathematically rigorous presentation that is directed toward several audiences. The practitioner can begin with the self-contained overview of our findings in the final chapter, and then refer to previous chapters as needed for details. Optimization modelers can make use of the linear and integer programming formulations we provide. Our proof techniques may be useful to fairness researchers who wish to obtain further results. Welfare economists may find of interest our marginal analyses of stakeholder incentives and cross-subsidies. Most of all, we hope to encourage and facilitate the current trend toward taking fairness seriously in decision models.

Özgün Elçi's contributions to this work were completed prior to joining Amazon.

Seattle, WA, USA — Özgün Elçi
Pittsburgh, PA, USA — John Hooker
Pittsburgh, PA, USA — Peter Zhang
November 2024

Contents

1 Introduction 1
1.1 The Fairness Modeling Challenge 1
1.2 Structural Analysis 2
1.3 Axiomatic and Bargaining Arguments 3
1.4 Plan of the Book 4
References 5

2 Optimizing with Fairness 9
2.1 Two Generic Models 9
2.2 A Modeling Example 10
2.3 A Simplified Generic Model 11
References 15

3 Hierarchical Distribution 17
3.1 Regional Decomposability 17
3.2 Monotonic Separability 19
References 19

4 Incentives and Sharing 21
4.1 Role of Marginal Analysis 21
4.2 Formulas for Marginal Analysis 22

5 Utilitarian and Maximin Criteria 23
5.1 Mathematical Formulation 23
5.2 Socially Optimal Distributions 24
5.3 Hierarchical Distributions 25
5.4 Incentives and Sharing 26
References 26

6 The Leximax Criterion ... 27
6.1 Mathematical Formulation ... 27
6.2 Socially Optimal Distribution ... 29
6.3 Hierarchical Distributions ... 30
6.4 Incentives and Sharing ... 31
Reference ... 31

7 Beta Fairness ... 33
7.1 Mathematical Formulation ... 33
7.2 Socially Optimal Solution ... 34
7.3 Hierarchical Distribution ... 36
7.4 Incentives and Sharing ... 37
References ... 37

8 Alpha Fairness ... 39
8.1 Mathematical Formulation ... 39
8.2 Socially Optimal Distribution ... 42
8.3 Hierarchical Distribution ... 44
8.4 Incentives and Sharing ... 44
References ... 44

9 Kalai–Smorodinsky Bargaining ... 47
9.1 Mathematical Formulation ... 47
9.2 Socially Optimal Distribution ... 48
9.3 Hierarchical Distribution ... 50
9.4 Incentives and Sharing ... 51
References ... 51

10 Utility Threshold Criterion ... 53
10.1 Mathematical Formulation ... 53
10.2 Socially Optimal Distribution ... 56
10.3 Hierarchical Distribution ... 59
10.4 Incentives and Sharing ... 61
References ... 61

11 Equity Threshold Criterion ... 63
11.1 Mathematical Formulation ... 63
11.2 Socially Optimal Distribution ... 65
11.3 Hierarchical Distribution ... 69
11.4 Incentives and Sharing ... 69
Reference ... 71

12 Utility Threshold Criterion with Leximax Fairness 73
12.1 Mathematical Formulation 73
12.2 Socially Optimal Distribution 74
12.3 Hierarchical Distribution 81
12.4 Incentives and Sharing 81
Reference 82

13 Inequality Bounds: Range 83
13.1 Mathematical Formulation 83
13.2 Socially Optimal Distribution 84
13.3 Hierarchical Distribution 85
13.4 Incentives and Sharing 86

14 Inequality Bounds: Gini Coefficient 87
14.1 Mathematical Formulation 87
14.2 Socially Optimal Solution 88
14.3 Hierarchical Distribution 91
14.4 Incentives and Sharing 92

15 Conclusions 95
15.1 Balancing Equity and Efficiency 95
15.2 Socially Optimal Distributions 96
15.3 Hierarchical Distributions 100
15.4 Incentives and Sharing 102
15.5 A Final Observation 104

Trade-Off Plots for Two Stakeholders 105

1 Introduction

1.1 The Fairness Modeling Challenge

There is growing interest in incorporating equity and fairness into optimization models. This concern arises in a number of application areas, including healthcare, disaster relief, facility location, task assignment, telecommunications, and machine learning. The aim is to maximize total benefit or minimize total cost, but only while ensuring that benefits and costs are equitably distributed among the relevant stakeholders.

For example, if donated organs are allocated in the most economically efficient fashion, patients with certain medical conditions may wait far longer for a transplant than other patients [1]. If earthquake shelters are located so as to minimize average distance from residents, persons living in less densely populated areas may have much further to travel [2]. If a machine learning algorithm awards mortgage loans so as to maximize expected earnings, members of a minority group may find themselves unable to obtain loans even when they are financially responsible [3]. If traffic signals at intersections are timed to maximize traffic throughput, motorists on side streets may have to wait forever for a green light [4].

The incorporation of fairness into an optimization model raises the issue of how it should be formulated mathematically. One attractive mechanism for doing so is to introduce a *social welfare function* (SWF) that reflects the desirability of a given distribution of benefits or costs. The SWF is a function of a utility vector, where each component of the vector represents the utility enjoyed by an individual stakeholder. Utility can be any relevant type of benefit, such as profit, health, or negative cost. The welfare economics literature [5] distinguishes *ordinal* SWFs, which rank utility vectors by desirability, from *cardinal* SWFs, which assign numerical values to utility vectors, with more desirable distributions receiving larger values. We study cardinal SWFs because they can serve as the objective function of an optimization model. The modeler can select a SWF that strikes an appropriate balance between total utility and fairness, and then maximize the SWF.

Ö. Elçi et al., *The Structure of Fair Solutions*, Synthesis Lectures on Operations Research and Applications, https://doi.org/10.1007/978-3-031-82190-5_1

A wide range of SWFs have been proposed for optimization models, many of them surveyed in [6, 7]. In addition, various SWFs have been compared in the context of particular application domains [8–12]. Given this plethora of SWFs, a modeler must select one with characteristics that are suitable for a given application. The aim of this book is to help inform this selection process. It does so by *characterizing the optimal solutions that SWFs actually deliver*.

1.2 Structural Analysis

Our primary research objective is to understand the structural properties of socially optimal distributions that result from various social welfare functions. This differs from most previous work in the fairness area. Historically, several SWFs have been characterized *ex ante* by showing that they can be derived from certain axioms, or defended as the outcome of certain bargaining procedures. These derivations are theoretically interesting and are briefly summarized in the next section. However, they tend to rely on strong assumptions that are implausible or difficult to assess in practice. We therefore propose that SWFs be characterized and evaluated *ex post* by examining the mathematical properties of the solutions they produce. This type of analysis is not only more relevant to practice, but it reveals properties that are interesting and useful in their own right. Very few of the structural results we derive have, to our knowledge, been observed elsewhere.

A number of previous studies could be viewed as *ex post* analysis in a general sense. For example, several examine the trade-off between overall utility and fairness, often referred to as the "price of fairness" [13–19]. These results can be informative when maximizing total utility subject to a fairness constraint, or vice versa. Yet the modeler must somehow determine how high a price is acceptable, a decision that is exogenous to the optimization model rather than determined by an SWF-based objective function. Also these studies generally do not describe structural properties of solutions obtained from maximizing an SWF.

There is also a vast literature on group fairness in AI-based decision-making, much of it surveyed in [20–23]. It investigates various parity metrics that are used for assessing or reducing bias in machine learning. Some of this literature conducts *ex post* analysis, normally empirical in nature and based on particular applications, but the accuracy/fairness trade-off is again exogenous. However, AI researchers have shown some recent interest in optimizing welfare-oriented fairness measures [9, 24, 25]. Since maximizing a SWF can achieve welfare-sensitive parity for many groups simultaneously, without having to identify them in advance [26], our structural results could be useful in a machine learning context.

Our *ex post* analysis, then, is distinguished by its focus on the structural properties of socially optimal solutions. It has three prongs that, together, we believe may assist in the choice of a suitable SWF for a particular purpose.

- *Structural characteristics in the presence of resource constraints.* For each of several SWFs, we describe the mathematical structure of a solution that results from maximizing that function subject to generic resource constraints. This may reveal properties that indicate whether the SWF is a suitable choice for a particular application domain. The constraints consist of a simple budget constraint and, optionally, bounds on the resources allotted to each stakeholder. This minimal constraint set is entirely unrealistic for most applications, but this is by design. Maximizing social welfare subject to a complex, realistic constraint set may result in a solution that is reasonable only because the *constraints* happen to exclude unjust solutions, and not because the model recognizes these solutions as unjust. In effect, realistically complex constraints can conceal the inherent inadequacy of the objective function when simpler constraints would lay bare its true nature.
- *Structure of hierarchical distributions.* We also examine the structure of solutions in a hierarchical distribution network. This represents the common situation in which a national authority or other central administration allocates resources to regions, which in turn combine these with their own resources for distribution to its subregions or institutions. Somewhat surprisingly, we find that for some SWFs, a region's distribution plan for subregions differs from that anticipated by the central administration, even when it uses the same SWF. In such cases, we say the national solution is not *regionally decomposable*. The presence or absence of this property may be an important consideration for whether a SWF is selected.
- *Incentives and sharing.* Finally, we ask whether a given SWF provides an incentive for stakeholders to use their allotted resources more efficiently, and whether doing so benefits others or pits one in competition against them. The choice of SWF can therefore incentivize either cooperation or competition. We address these matters with a marginal analysis of benefits or costs to the stakeholders as a function of a given stakeholder's efficiency improvements.

1.3 Axiomatic and Bargaining Arguments

Several social welfare functions discussed in this book have been characterized *ex ante* by showing that they can be derived from certain axioms or bargaining procedures. We very briefly describe these derivations, to clarify the contrast with *ex post* analysis. Useful expositions of classical results in this area can be found in [27–30].

Social welfare functions that combine equity and efficiency comprise one of the oldest research topics in social choice theory, beginning at least as early as John Nash's 1950 axiomatic derivation of proportional fairness [31]. His axioms include anonymity (stakeholders can be interchanged without affecting social welfare), a Pareto axiom, and the independence of irrelevant alternatives. This last axiom requires that if a given distribution of utilities is socially optimal, then it remains optimal if the feasible set is reduced without excluding it. There is also an assumption of *scale invariance*, meaning that each stakeholders's utility measure can be rescaled by a different factor without changing how the SWF ranks utility vectors. This implies that utility levels can be compared across stakeholders only in a limited sense.

If the independence axiom is replaced by a monotonicity axiom, one can derive the bargaining solution for two stakeholders studied by Kalai and Smorodinsky [30, 32]. Monotonicity means that enlarging the set of feasible utility vectors in a certain way does not reduce the utility allotment to any stakeholder. Similar axioms imply a purely utilitarian SWF when combined with *unit comparability*, and a maximin/minimax SWF when combined with *level comparability* [27, 28]. These derivations likewise rest on strong assumptions regarding the lack of comparability of utilities across stakeholders.

The classical results are often difficult to apply, because interpersonal comparability in practice is typically much greater than they assume. Utility often takes the form of profit or negative cost, which can be compared on an absolute scale across stakeholders. Even such health-related utility measures as quality-adjusted life years (QALYs) have a great degree of interpersonal comparability. A more recent axiomatic result [33, 34] does not explicitly rely on an interpersonal noncomparability assumption. It states that a SWF satisfying certain axioms must belong to a certain broad family of functions, including alpha fairness as well as maximin and minimax criteria. However, while this is an interesting and comprehensive result, it relies essentially on an axiom of partition that is rather abstract and difficult to assess in a practical context. It basically says that the ratio of social welfare values of two distributions must be the geometric or power mean of the ratios that result when the stakeholders are partitioned into subgroups.

Bargaining justifications have also been given for certain SWFs. For example, it is shown in [35–37] that the Nash solution is the (asymptotic) outcome of certain rational bargaining procedures. Similar arguments can be advanced for the Kalai–Smorodinsky bargaining solution. These results are helpful only if one is willing to accept a bargaining solution regardless of what distributional characteristics it might have. It is a fundamental question of political economy whether a rational negotiation process (one form of "procedural justice") necessarily produces a just outcome ("substantive justice") [38–41].

1.4 Plan of the Book

After this introductory chapter, three chapters set the stage for the main results that follow. Chapter 2 shows how an optimization model can incorporate fairness in the form of a social welfare function, and presents an illustrative healthcare example. It then specifies the generic resource constraints with respect to which most of the structural results in the book are derived. Chapter 3 formulates the optimization problem on a hierarchical network and defines concepts of collapsibility, monotone separability, and regional decomposability. Chapter 4 explains how we analyze incentives and sharing.

The bulk of the book consists of ten chapters that examine a series of increasingly sophisticated SWFs. They include utilitarian, Rawlsian maximin, and leximax SWFs, as well as beta fairness, proportional fairness (the Nash bargaining solution), its generalization to alpha fairness, the Kalai–Smorodinsky bargaining solution, and recently proposed threshold functions that combine utilitarian with maximin or leximax fairness. The focus is primarily on SWFs that combine equity with utilitarian efficiency in some fashion, because efficiency is as important as equity in most practical applications. However, the last two chapters in this series study problems that maximize total utility subject to limits on inequality, due to the interesting facts they reveal: one problem is closely related to maximizing a threshold SWF, and the other has an intriguing solution structure that could help explain why so many societies have historically consisted of two or three major socioeconomic classes.

Each of these ten chapters contains four sections. The first contains a general description and mathematical formulation of the SWF, followed by a mathematical programming model that is suitable for practical solution. Most of these models are similar to those stated in [6]. The next section contains the main result of the chapter: a structural description of socially optimal solutions subject to generic resource constraints, along with an interpretation of the result. The theorems are elementary for the first few SWFs but become highly nontrivial for the later ones, a few requiring several pages of proof. The last two sections of each chapter contain an investigation of whether the problem is regionally decomposable and a marginal analysis of incentives and sharing.

The final chapter draws together the foregoing results to construct a general guide to selecting an appropriate SWF for a given application. In addition, several tables summarize the results in concise form.

References

1. McElfresh, C., Dickerson, J.: Balancing lexicographic fairness and a utilitarian objective with application to kidney exchange. In: Proceedings of AAAI Conference on Artificial Intelligence (AAAI 2018), pp. 1161–1168 (2018)
2. Mostajabdaveh, M., Gutjahr, W.J., Sibel Salman, F.: Inequity-averse shelter location for disaster preparedness. IISE Transactions **51**(8), 809–829 (2019)
3. Saxena, N.A., Huang, K., DeFilippis, E., Radanovic, G., Parkes, D.C., Liu, Y.: How do fairness definitions fare? Testing public attitudes towards three algorithmic definitions of fairness in loan allocations. Artificial Intelligence **283**, 1–15 (2020)
4. Chen, L.W., Sharma, P., Tseng, Y.C.: Dynamic traffic control with fairness and throughput optimization using vehicular communications. IEEE Journal on Selected Areas in Communications **31**, 504–512 (2013)
5. Adler, M.D.: Measuring Social Welfare. Oxford University Press (2019)
6. Chen, V., Hooker, J.N.: A guide to formulating fairness in an optimization model. Annals of Operations Research **326**, 581–619 (2023)
7. Karsu, O., Morton, A.: Inequity averse optimization in operational research. European Journal of Operational Research **245**, 343–359 (2015)

8. Colquitt, J.A., Rodell, J.B.: Measuring justice and fairness. In: R.S. Cropanzano, M.L. Ambrose (eds.) Oxford Handbook of Justice in the Workplace, pp. 187–202. Oxford University Press (2015)
9. Heidari, H., Ferrari, C., Gummadi, K., Krause, A.: Fairness behind a veil of ignorance: A welfare analysis for automated decision making. Advances in Neural Information Processing Systems **31**, 1265–1276 (2018)
10. Ogryczak, W., Luss, H., Pióro, M., Nace, D., Tomaszewski, A.: Fair optimization and networks: A survey. Journal of Applied Mathematics **2014**, 1–25 (2014)
11. Rahmattalabi, A., Jabbari, S., Lakkaraju, H., Vayanos, P., Izenberg, M., Brown, R., Rice, E., Tambe, M.: Fair influence maximization: A welfare optimization approach. In: Proceedings of the 35th AAAI Conference on Artificial Intelligence, pp. 11630–11638 (2021)
12. Schminke, M., Johnson, M., Rice, D.: Justice and organizational structure: A review. In: R.S. Cropanzano, M.L. Ambrose (eds.) Oxford Handbook of Justice in the Workplace, pp. 541–560. Oxford University Press (2015)
13. Bertsimas, D., Farias, V., Trichakis, N.: The price of fairness. Operations Research **59**, 17–31 (2011)
14. Bertsimas, D., Farias, V., Trichakis, N.: On the efficiency-fairness trade-off. Management Science **58**, 2234–2250 (2012)
15. Bertsimas, D., Papalexopoulos, T., Trichakis, N., Wang, Y., Hirose, R., Vagefi, P.A.: Balancing efficiency and fairness in liver transplant cccess: Tradeoff curves for the assessment of organ distribution policies. Transplantation **104**, 981–987 (2020)
16. Huang, M., Smilowitz, K., Balcik, B.: Models for relief routing: Equity, efficiency and efficacy. Transportation Research Part E **48**, 2–18 (2012)
17. Manshadi, V., Niazadeh, R., Rodilitz, S.: Fair dynamic rationing. In: Proceedings of the 22nd ACM Conference on Economics and Computation, pp. 694–695 (2021)
18. Nguyen, Q., Das, S., Garnett, R.: Scarce societal resource allocation and the price of (local) justice. In: Proceedings of the 35th AAAI Conference on Artificial Intelligence, pp. 5628–5636 (2021)
19. Sinclair, S.R., Banerjee, S., Yu, C.L.: Sequential fair allocation: Achieving the optimal envy-efficiency tradeoff curve. ACM SIGMETRICS Performance Evaluation Review **50**, 95–96 (2022)
20. Castelnovo, A., Crupi, R., Greco, G., Regoli, D., Penco, I.G., Cosentini, A.C.: A clarification of the nuances in the fairness metrics landscape. Scientific Reports **12**(1), 1–21 (2022)
21. Jo, N., Tang, B., Dullerud, K., Aghaei, S., Rice, E., Vayanos, P.: Fairness in contextual resource allocation systems: Metrics and incompatibility results. In: Proceedings of the 37th AAAI Conference on Artificial Intelligence, pp. 11837–11846 (2023)
22. Mehrabi, N., Morstatter, F., Saxena, N., Lerman, K., Galstyan, A.: A survey on bias and fairness in machine learning. ACM Computing Surveys **54**, 1–35 (2021)
23. Verma, S., Rubin, J.: Fairness definitions explained. In: Proceedings of the International Workshop on Software Fairness (FairWare), pp. 1–7 (2023)
24. Heidari, H., Loi, M., Gummadi, K.P., Krause, A.: A moral framework for understanding fair ML through economic models of equality of opportunity. In: Proceedings of the Conference on Fairness, Accountability, and Transparency, pp. 181–190 (2019)
25. Hu, L., Chen, Y.: Fair classification and social welfare. In: Proceedings of the Conference on Fairness, Accountability, and Transparency, pp. 535–545 (2020)
26. Chen, V., Hooker, J.N., Leben, D.: Assessing group fairness with social welfare optimization. In: CPAIOR Proceedings, pp. 208–225. Springer (2024)
27. Gaertner, W.: A Primer in Social Choice Theory. Oxford University Press (2009)
28. Roberts, K.W.S.: Interpersonal comparability and social choice theory. Review of Economic Studies **47**, 421–439 (1980)

29. Roemer, J.E.: Theories of Distributive Justice. Harvard University Press (1996)
30. Thompson, W.: Cooperative models of bargaining. In: R.J. Aumann, S. Hart (eds.) Handbook of Game Theory, vol. 2, pp. 1237–1284. North-Holland (1994)
31. Nash, J.: The bargaining problem. Econometrica **18**, 155–162 (1950)
32. Kalai, E., Smorodinsky, M.: Other solutions to Nash's bargaining problem. Econometrica **43**, 513–518 (1975)
33. Lan, T., Chiang, M.: An axiomatic theory of fairness in resource allocation. Tech. rep., Princeton University (2011)
34. Lan, T., Kao, D., Chiang, M.: An axiomatic theory of fairness in network resource allocation. In: IEEE INFOCOM (2010)
35. Binmore, K., Rubinstein, A., Wolinsky, A.: The Nash bargaining solution in economic modeling. RAND Journal of Economics **17**, 176–188 (1986)
36. Harsanyi, J.C.: Rational Behavior and Bargaining Equilibrium in Games and Social Situations. Cambridge University Press (1977)
37. Rubinstein, A.: Perfect equilibrium in a bargaining model. Econometrica **50**, 97–109 (1982)
38. Cropanzana, R., Bowen, D.E., Gilliland, S.W.: The management of organizational justice. Academy of Management Perspectives **21**, 34–48 (2007)
39. Nozick, R.: Anarchy, State, and Utopia. Blackwell (1974)
40. Rawls, J.: A Theory of Justice (revised). Harvard University Press, original edition 1971 (1999)
41. Scanlon, T.M.: The diversity of objections to inequality. In: T.M. Scanlon (ed.) The Difficulty of Tolerance: Essays in Political Philosophy, pp. 202–218. Cambridge University Press (2003)

Optimizing with Fairness

2

2.1 Two Generic Models

The aim of this book is to explore options for incorporating fairness into an optimization model. We assume that the variables in the model are of two types: variables $\boldsymbol{u} = (u_1, \ldots, u_n)$ that represent the utility u_i allocated to each stakeholder i, and the remaining decision variables $\boldsymbol{x} = (x_1, \ldots, x_m)$ of the model. As noted earlier, utility can be any type of benefit, such as profit, goods, services, health outcomes, or negative cost. Stakeholders can be individuals, groups of individuals, organizations, geographical regions, or any other set of utility recipients for which fairness is an issue. We represent the constraints of the model by writing $(\boldsymbol{u}, \boldsymbol{x}) \in S$, where S is the feasible set. These include constraints that specify how $\boldsymbol{u}$ depends on $\boldsymbol{x}$.

We consider two methods of injecting fairness into an optimization model. For the most part, we examine models that maximize social welfare subject to constraints, where social welfare is measured by a function $W(\boldsymbol{u})$. The social welfare function $W(\boldsymbol{u})$ can reflect the overall utility level as well as the degree to which the utilities $\boldsymbol{u} = (u_1, \ldots, u_n)$ are distributed fairly in some sense. Models of this kind have the form:

$$\max_{\boldsymbol{u},\boldsymbol{x}} \left\{ W(\boldsymbol{u}) \;\middle|\; (\boldsymbol{u}, \boldsymbol{x}) \in S \right\} \tag{2.1}$$

We also give some attention to models that maximize total utility subject to an upper bound on some measure $I(\boldsymbol{u})$ of inequality. They have the form:

$$\max_{\boldsymbol{u},\boldsymbol{x}} \left\{ \sum_i u_i \;\middle|\; I(\boldsymbol{u}) \leqslant \text{UB},\ (\boldsymbol{u}, \boldsymbol{x}) \in S \right\} \tag{2.2}$$

where UB is the desired upper bound.

Ö. Elçi et al., *The Structure of Fair Solutions*, Synthesis Lectures on Operations Research and Applications, https://doi.org/10.1007/978-3-031-82190-5_2

We are primarily concerned with models that maximize social welfare, because models (2.2) that constrain inequality rely on an exogenous choice of the upper bound UB. There is no guidance in the model as to how one should balance total utility with fairness. Model (2.1), on the other hand, allows one to select a social welfare function that balances total benefit and fairness in a particular fashion that is suitable for a given application. Most of the SWFs we consider are explicitly designed to strike such a balance, and some have an interpretable parameter that adjusts the relative importance of fairness. In addition, a SWF may have theoretical justification, or its solutions may have structural properties that argue for its application in a particular instance. Indeed, the main purpose of this book is to discover such structural properties.

This said, we investigate two inequality-constrained models in Chaps. 13 and 14 because they have interesting and even surprising solutions. These models maximize total utility subject to a bound on the range of utilities in one case, and an unnormalized Gini coefficient in the other.

Table 2.1 defines notation that is used throughout the book.

2.2 A Modeling Example

We can illustrate the generic models (2.1) and (2.2) with a medical provision problem similar to that described in [1]. The objective is to allocate healthcare resources in a manner that is both equitable and efficient, subject to a budget limitation. The stakeholders are groups of patients, where the patients in each group are diagnosed with a similar condition. We assume each group i consists of n_i patients. A group's utility is measured by the total quality-adjusted life years (QALYs) that patients in the group will experience. The problem is to decide what fraction of each group is to receive a specified treatment.

We suppose p_i is the cost per patient of administering treatment in group i, q_i a patient's average net gain in QALYs when the treatment is administered, and r_i the QALYs that result from medical management without the treatment. Variables x_i indicate the fraction of patients in group i that are provided treatment. In the event of partial funding for a group, medical personnel make triage decisions based on individual situations. The expected average utility u_i per patient in each group i (measured in QALYs) is estimated by an affine function $q_i x_i + r_i$, on the assumption that the additional expected utility resulting from treatment is roughly proportional to the number of patients treated. The resulting welfare maximizing model is

$$\max_{\boldsymbol{u},\boldsymbol{x}} \left\{ W(\boldsymbol{u}) \;\middle|\; \sum_i n_i p_i x_i \leqslant \tilde{B};\;\; u_i = q_i x_i + r_i,\; 0 \leqslant x_i \leqslant 1,\; \text{all } i \right\}$$

where $\tilde{B}$ is the available budget. The SWF $W(\boldsymbol{u})$ would presumably take into account the group size so that total QALYs can be balanced against the distribution $\boldsymbol{u}$ of average QALYs. The problem is solved in [1] with binary x_i using a utility threshold SWF with maximin

Table 2.1 List of symbols

u_i	Utility assigned to stakeholder i
$\boldsymbol{u}$	Vector $(u_1, \ldots, u_n)$ of utilities
$\boldsymbol{u}(I)$	Vector of utilities u_i for $i \in I$
$u_{\min}$	$\min_i \{u_i\}$
$W(\boldsymbol{u})$	Social welfare function W applied to vector $\boldsymbol{u}$
SWF	Social welfare function
c_i	Lower bound on utility u_i
d_i	Upper bound on utility u_i
B	Upper bound on available resources
B_k	Upper bound on available resources in region k
a_i	Resource cost of creating one unit of utility for stakeholder i
$\boldsymbol{a}$	Vector $(a_1, \ldots, a_n)$ of costs
$a(j)$	$\sum_{i=1}^{j} a_i$
$a(j\!:\!k)$	$\sum_{i=j}^{k} a_i$
$a(I)$	$\sum_{i \in I} a_i$
α	Equity/efficiency trade-off parameter for alpha fairness
β	Equity/efficiency trade-off parameter for beta fairness
Δ	Equity/efficiency trade-off parameter for threshold SWFs
R	Upper bound on utility range
D	Upper bound on unnormalized Gini coefficient
$\boldsymbol{e}$	Vector of ones
$\boldsymbol{e}_i$	ith unit vector $(0, \ldots, 1, \ldots, 0)$

(described in Chap. 10), and is solved in [2] with a threshold SWF with leximax (Chap. 12). Other SWFs may be appropriate as well.

An inequality-constrained model of this problem can also be used. For example, it could maximize total QALYs subject to a bound UB on some inequality measure $I(\boldsymbol{u})$ and have the form:

$$\max_{\boldsymbol{u},\boldsymbol{x}} \left\{ \sum_i n_i u_i \;\middle|\; I(\boldsymbol{u}) \leqslant \text{UB};\ \sum_i n_i p_i x_i \leqslant \tilde{B};\ \ u_i = q_i x_i + r_i,\ 0 \leqslant x_i \leqslant 1,\ \text{all } i \right\}$$

2.3 A Simplified Generic Model

For purposes of studying the structure of optimal solutions, we focus attention on a model that maximizes social welfare subject to generic resource constraints. As explained in the previous chapter, a simple model of this kind exposes the fairness properties of an SWF without the confounding effect of a complex constraint set.

In this simplified model, a stakeholder's utility u_i is assumed to be a linear function $(1/a_i)x_i$ of the resource x_i allocated to the stakeholder, where $a_i > 0$. The coefficient $1/a_i$ is the *conversion efficiency* of stakeholder i, using terminology from Sen's capabilities approach [3]. Thus a_i is the resource cost of generating one unit of utility for stakeholder i. A large value of a_i indicates that it is expensive to improve the stakeholders welfare, perhaps due to a disease that is costly to treat. There are also a budget constraint and, optionally, bounds on the resource allocated to each stakeholder. The resulting model is

$$\max_{\boldsymbol{u},\boldsymbol{x}} \left\{ W(\boldsymbol{u}) \;\middle|\; \sum_i x_i \leqslant B,\ \bar{\boldsymbol{c}} \leqslant \boldsymbol{x} \leqslant \bar{\boldsymbol{d}};\ \ u_i = x_i/a_i,\ \text{all } i \right\} \tag{2.3}$$

where the budget B limits the total amount of resource to be distributed.

Due to the equations $u_i = x_i/a_i$, we can eliminate the variables x_i and write (2.3) as

$$\max_{\boldsymbol{u}} \left\{ W(\boldsymbol{u}) \mid \boldsymbol{a}^{\mathsf{T}}\boldsymbol{u} \leqslant B,\ \boldsymbol{c} \leqslant \boldsymbol{u} \leqslant \boldsymbol{d} \right\} \tag{2.4}$$

where $c_i = \bar{c}_i/a_i$ and $d_i = \bar{d}_i/a_i$ for all i. We will refer to problem (2.4) as maximizing social welfare *subject to a budget constraint and utility bounds*. The upper bounds d_i can reflect the fact that greater resources can yield greater utility only up to a point; once the disease is cured, there is no need to provide more medical resources.

The lower bounds c_i can impose a floor on the welfare of each stakeholder, or they may reflect pre-existing stakeholder utilities before the distribution problem is solved. The latter situation occurs when one is distributing incremental utility but wishes to take into account total utility when assessing fairness. For example, if one is designing tax policy, it may be important to consider the effect of tax breaks on the utility of the taxpayer's overall financial situation. Then u_i would represent taxpayer i's utility after the tax break, and c_i the taxpayer's utility before the tax break. On the other hand, when locating transit stops, it is perhaps more appropriate and practical for a fairness assessment to consider only the utility u_i added by locating a stop near a given neighborhood i. In this case, u_i is the incremental utility and $c_i = 0$.

We often investigate the behavior of a SWF without upper and/or lower bounds on the utilities, because this can drastically simplify the mathematical analysis while still delivering interesting and useful insights. The solutions of bounded problems are often qualitatively similar to those obtained without bounds, and our analysis indicates the proof strategy one would use to obtain more general results.

The social welfare optimization problem (2.4) without positive lower bounds ($c_i = 0$ for all i) can be written as

$$\max_{\boldsymbol{u}} \left\{ W(\boldsymbol{u}) \mid \boldsymbol{a}^{\mathsf{T}}\boldsymbol{u} \leqslant B,\ \boldsymbol{0} \leqslant \boldsymbol{u} \leqslant \boldsymbol{d} \right\} \tag{2.5}$$

We refer to this problem as maximizing social welfare *subject to a budget constraint and upper bounds*. Solutions of this model frequently assign several stakeholders utilities $u_i = 0$, but this does not necessarily mean that these stakeholders are left with no utility at all. For example, if u_i represents the incremental utility distributed to stakeholder i, then $u_i = 0$ means that stakeholder i's utility remains at a pre-existing level that is not part of the fairness assessment.

If there are no lower bounds ($c_i = 0$ for all i) and no nonredundant upper bounds ($d_i \geqslant B/a_i$ for all i), model (2.4) becomes

$$\max_{\boldsymbol{u}} \left\{ W(\boldsymbol{u}) \mid \boldsymbol{a}^{\mathsf{T}}\boldsymbol{u} \leqslant B,\ \boldsymbol{u} \geqslant \boldsymbol{0} \right\} \tag{2.6}$$

which we refer to as maximizing social welfare *subject to a budget constraint.*

One could in principle introduce a nonlinear utility function into the model (2.3), which would result in a nonlinear budget constraint in (2.4)–(2.6). For example, one may wish to represent decreasing returns to scale with a strictly concave utility function, such as $u_i = x_i^{\gamma}/a_i$ for $0 < \gamma < 1$. This results in the nonlinear budget constraint:

$$\sum_i a_i^{1/\gamma} u_i^{1/\gamma} \leqslant B \tag{2.7}$$

Yet the choice of which nonlinear utility function to use would strongly influence the equity properties of the solution. A linear utility function allows us to account for conversion efficiencies while reflecting the choice of fairness priorities as transparently as possible in the social welfare function, rather than embedding them in the constraint set.

For example, a problem that maximizes $W(\boldsymbol{u})$ subject to (2.7) can be reformulated, using a simple change of variable $u_i = \bar{u}_i^{\gamma}$, as maximizing the revised SWF

$$\overline{W}(\bar{\boldsymbol{u}}) = W(\bar{u}_1^{\gamma}, \ldots, \bar{u}_n^{\gamma})$$

subject to the linear budget constraint

$$\sum_i a_i^{1/\gamma} \bar{u}_i \leqslant B \tag{2.8}$$

This moves all elements of the problem that affect fairness into the SWF. An important special case of this problem arises when $W(\boldsymbol{u})$ is the purely utilitarian criterion $\sum_i u_i$. In this case, the problem becomes one of maximizing alpha fairness subject to (2.8), as we will see in Chap. 8.

The simplified model (2.4) is not only useful for exploring the structure of optimal solutions, but is itself adequate to formulate some important applications. For example, the medical provision problem described in the previous section has the form (2.4) with

$$B = \tilde{B} + \sum_i \frac{n_i p_i r_i}{q_i};\quad a_i = \frac{n_i p_i}{q_i},\ c_i = r_i,\ d_i = q_i + r_i,\ \text{all } i$$

As a preview of how SWFs can differ, Table 2.2 compares the solutions they yield in a very simple example. There are eight stakeholders, and the SWF is maximized subject

Table 2.2 Optimal utility (and resource) distributions for various SFWs subject to a budget constraint $\boldsymbol{au} \leqslant B$, where $\boldsymbol{a} = (5, 6, \ldots, 12)$ and $B = 100$

SWF	Stakeholders															
	1		2		3		4		5		6		7		8	
Utilitarian	**20.0**	(100)	**0.0**	(0)	**0.0**	(0)	**0.0**	(0)	**0.0**	(0)	**0.0**	(0)	**0.0**	(0)	**0.0**	(0)
Maximin	**1.5**	(7)	**1.5**	(9)	**1.5**	(10)	**1.5**	(12)	**1.5**	(13)	**1.5**	(15)	**1.5**	(16)	**1.5**	(18)
Beta fairness*																
$\beta = 1$	**20.0**	(100)	**0.0**	(0)	**0.0**	(0)	**0.0**	(0)	**0.0**	(0)	**0.0**	(0)	**0.0**	(0)	**0.0**	(0)
$\beta < 1$	**1.5**	(7)	**1.5**	(9)	**1.5**	(10)	**1.5**	(12)	**1.5**	(13)	**1.5**	(15)	**1.5**	(16)	**1.5**	(18)
Alpha fairness																
$\alpha = 0.5$	**3.9**	(20)	**2.7**	(16)	**2.0**	(14)	**1.5**	(12)	**1.2**	(11)	**1.0**	(10)	**0.8**	(9)	**0.7**	(8)
$\alpha = 1.0$	**2.5**	(13)	**2.1**	(13)	**1.8**	(13)	**1.6**	(13)	**1.4**	(13)	**1.3**	(13)	**1.1**	(13)	**1.0**	(13)
$\alpha = 2.0$	**1.9**	(10)	**1.8**	(11)	**1.6**	(11)	**1.5**	(12)	**1.4**	(13)	**1.4**	(14)	**1.3**	(14)	**1.2**	(15)
Kalai–Smor.	**2.5**	(13)	**2.1**	(13)	**1.8**	(13)	**1.6**	(13)	**1.4**	(13)	**1.3**	(13)	**1.1**	(13)	**1.0**	(13)
Utility threshold with maximin																
$\Delta \leqslant 8.235$	**20.0**	(100)	**0.0**	(0)	**0.0**	(0)	**0.0**	(0)	**0.0**	(0)	**0.0**	(0)	**0.0**	(0)	**0.0**	(0)
$\Delta \geqslant 8.235$	**1.5**	(7)	**1.5**	(9)	**1.5**	(10)	**1.5**	(12)	**1.5**	(13)	**1.5**	(15)	**1.5**	(16)	**1.5**	(18)
Equity threshold with maximin																
$\Delta = 1$	**2.1**	(10)	**2.1**	(13)	**2.1**	(15)	**2.1**	(17)	**1.1**	(10)	**1.1**	(11)	**1.1**	(12)	**1.1**	(13)
$\Delta = 3$	**3.3**	(17)	**3.3**	(20)	**3.3**	(23)	**3.3**	(27)	**0.3**	(3)	**0.3**	(3)	**0.3**	(4)	**0.3**	(4)
$\Delta = 4$	**4.0**	(20)	**4.0**	(24)	**4.0**	(28)	**3.5**	(28)	**0.0**	(0)	**0.0**	(0)	**0.0**	(0)	**0.0**	(0)
$\Delta = 6$	**6.0**	(30)	**6.0**	(36)	**4.9**	(34)	**0.0**	(0)	**0.0**	(0)	**0.0**	(0)	**0.0**	(0)	**0.0**	(0)
Utility threshold with leximax																
$\Delta = 2$	**9.1**	(45)	**9.1**	(55)	**0.0**	(0)	**0.0**	(0)	**0.0**	(0)	**0.0**	(0)	**0.0**	(0)	**0.0**	(0)
$\Delta = 5$	**3.8**	(19)	**3.8**	(23)	**3.8**	(27)	**3.8**	(31)	**0.0**	(0)	**0.0**	(0)	**0.0**	(0)	**0.0**	(0)
$\Delta = 8$	**1.8**	(9)	**1.8**	(11)	**1.8**	(13)	**1.8**	(14)	**1.8**	(16)	**1.8**	(18)	**1.8**	(20)	**0.0**	(0)
Max utility subject to range bound																
$R = 1$	**2.1**	(10)	**2.1**	(13)	**2.1**	(15)	**2.1**	(17)	**1.1**	(10)	**1.1**	(11)	**1.1**	(12)	**1.1**	(13)
$R = 3$	**3.3**	(17)	**3.3**	(20)	**3.3**	(23)	**3.3**	(27)	**0.3**	(3)	**0.3**	(3)	**0.3**	(4)	**0.3**	(4)
$R = 4$	**4.0**	(20)	**4.0**	(24)	**4.0**	(28)	**3.5**	(28)	**0.0**	(0)	**0.0**	(0)	**0.0**	(0)	**0.0**	(0)
$R = 6$	**6.0**	(30)	**6.0**	(36)	**4.9**	(34)	**0.0**	(0)	**0.0**	(0)	**0.0**	(0)	**0.0**	(0)	**0.0**	(0)
Max utility subject to unnormalized Gini bound																
$G = 0.5$	**2.5**	(13)	**2.5**	(15)	**2.5**	(18)	**2.5**	(20)	**2.5**	(23)	**0.4**	(4)	**0.4**	(4)	**0.4**	(4)
$G = 1.0$	**4.8**	(24)	**4.8**	(29)	**1.9**	(14)	**1.9**	(16)	**1.9**	(18)	**0.0**	(3)	**0.0**	(4)	**0.0**	(4)
$G = 1.5$	**7.9**	(39)	**7.9**	(47)	**0.6**	(4)	**0.6**	(5)	**0.6**	(5)	**0.0**	(0)	**0.0**	(0)	**0.0**	(0)
$G = 2.0$	**15.8**	(79)	**3.5**	(21)	**0.0**	(0)	**0.0**	(0)	**0.0**	(0)	**0.0**	(0)	**0.0**	(0)	**0.0**	(0)

*β must be an integral multiple of $1/n$, where n is the number of stakeholders

to a budget constraint. The total resource budget is $B = 100$, and the cost coefficients are $(a_1, \ldots, a_8) = (5, 6, \ldots, 12)$. Note that the socially optimal utility distribution $\boldsymbol{u}$ is shown in boldface, while the corresponding resource distribution appears in parentheses. The numerical results in the table will be explained as we consider each of the SWFs.

References

1. Hooker, J.N., Williams, H.P.: Combining equity and utilitarianism in a mathematical programming model. Management Science **58**, 1682–1693 (2012)
2. Chen, V., Hooker, J.N.: Combining leximax fairness and efficiency in an optimization model. European Journal of Operational Research **299**, 235–248 (2022)
3. Sen, A.: Inequality Reexamined. Oxford University Press (1995)

Hierarchical Distribution 3

3.1 Regional Decomposability

Resources are frequently distributed in a hierarchical fashion [1]. For example, a national authority may allocate resources to regions, whereupon each region allocates its share (perhaps augmented by local funds) to subregions. The problem can be formulated as a single instance of the generic model (2.1) in which the national authority decides how much each region allocates to each of its subregions. This obviously determines the allocation to each region. The hierarchical problem is nonetheless worthy of special attention, because it can lead to somewhat paradoxical outcomes, depending on the mathematical properties of the SWF. If a region allocates its share of resources optimally to its subregions, the allocation may be *different than in the national solution*—even when it is based on the same SWF. We say that the distribution problem is *regionally decomposable* if this cannot occur for a given SWF. In this case, the region's allocation to subregions necessarily matches the allocation in the national solution.

Regional decomposability can have a bearing on the choice of SWF for a given application. If the problem is not regionally decomposable, the regional authorities may be unable to confirm the nationally recommended allocation and may therefore question it. In such cases, it may be desirable to switch to another SWF. If no other SWF is suitable for the application, the regional authorities must be persuaded that fairness within a region can be adequately assessed only from a national perspective. It is therefore important that practitioners be alerted to any lack of regional decomposability. For this reason, we examine each SWF discussed in ensuing chapters for regional decomposability, by either proving the property obtains or exhibiting a counterexample. To simplify discussion, we obtain these results for problems with a linear budget constraint. Any SWF that lacks decomposability in this simplified model obviously lacks it in general.

Ö. Elçi et al., *The Structure of Fair Solutions*, Synthesis Lectures on Operations Research and Applications, https://doi.org/10.1007/978-3-031-82190-5_3

To formulate the hierarchical problem, we suppose that each region k has a pre-existing resource budget B_k, and the national government must decide how much additional resource y_k to allocate to region k. If there are r regions, the generic model (2.1) becomes

$$\max_{\boldsymbol{u},\boldsymbol{y}} \left\{ W(\boldsymbol{u}) \,\middle|\, \begin{array}{l} \boldsymbol{e}^{\mathsf{T}}\boldsymbol{y} \leqslant B,\ \boldsymbol{c} \leqslant \boldsymbol{u} \leqslant \boldsymbol{d},\ \boldsymbol{y} \geqslant \boldsymbol{0} \\ \boldsymbol{a}^k \boldsymbol{u}^k \leqslant B_k + y_k,\ k = 1, \ldots, r \end{array} \right\} \tag{3.1}$$

where $\boldsymbol{e} = (1, \ldots, 1)$ and vector $\boldsymbol{u}^k$ contains the utilities of subregions of region k.

Interestingly, if we drop the requirement $\boldsymbol{y} \geqslant \boldsymbol{0}$ (i.e., we allow the national government to take resources from the regions), the model collapses into a single-level problem:

$$\max_{\boldsymbol{u}} \left\{ W(\boldsymbol{u}) \,\middle|\, \boldsymbol{a}^{\mathsf{T}}\boldsymbol{u} \leqslant B + \sum_{k=1}^{r} B_k,\ \boldsymbol{c} \leqslant \boldsymbol{u} \leqslant \boldsymbol{d} \right\} \tag{3.2}$$

We will say that a hierarchical problem is *collapsible* if it can be solved by solving its collapsed version (3.2). A problem is collapsible if each region's allocation in the collapsed problem (3.2) is no less than its stock already on hand.

Proposition 3.1 *A hierarchical problem* (3.1) *is collapsible if for any optimal solution* $\bar{\boldsymbol{u}}$ *of* (3.2), $\boldsymbol{a}^k\bar{\boldsymbol{u}}^k \geqslant B_k$ *for* $k = 1, \ldots, r$.

Proof Let $\bar{\boldsymbol{u}}$ be an optimal solution of (3.2) that satisfies $\boldsymbol{a}^k\bar{\boldsymbol{u}}^k \geqslant B_k$ for $k = 1, \ldots, r$. If we let $\bar{y}_k = \boldsymbol{a}^k\bar{\boldsymbol{u}}^k - B_k \geqslant 0$, then $(\bar{\boldsymbol{y}}, \bar{\boldsymbol{u}})$ is feasible in (3.1) because

$$\sum_{k=1}^{r} \bar{y}_k = \sum_{k=1}^{r} \boldsymbol{a}^k\bar{\boldsymbol{u}}^k - B_k \leqslant B \tag{3.3}$$

where the inequality is due to the constraint in (3.2). Also any $\boldsymbol{u}$ feasible in (3.1) is feasible in (3.2), which implies $W(\boldsymbol{u}) \leqslant W(\bar{\boldsymbol{u}})$ since $\bar{\boldsymbol{u}}$ is optimal in (3.2). Thus $(\bar{\boldsymbol{y}}, \bar{\boldsymbol{u}})$ is optimal in (3.1). □

The distribution problem is *regionally decomposable* when the optimal distribution within any region, given the resource subsidy provided by the national solution, is the same as in the national solution. To define this more precisely, consider the regional distribution problems:

$$\max_{\boldsymbol{u}^k} \left\{ W(\boldsymbol{u}^k) \mid \boldsymbol{a}^k\boldsymbol{u}^k \leqslant B_k + \bar{y}_k,\ \boldsymbol{c}^k \leqslant \boldsymbol{u}^k \leqslant \boldsymbol{d}^k \right\},\quad k = 1, \ldots, r \tag{3.4}$$

The hierarchical problem (3.1) is regionally decomposable if, given any optimal solution $(\bar{\boldsymbol{u}}, \bar{\boldsymbol{y}})$ of (3.1), and given any set of solutions $\hat{\boldsymbol{u}}^1, \ldots, \hat{\boldsymbol{u}}^r$ that are, respectively, optimal in the regional distribution problems (3.4), the solution $(\boldsymbol{u}, \boldsymbol{y}) = (\hat{\boldsymbol{u}}^1, \ldots, \hat{\boldsymbol{u}}^r, \bar{\boldsymbol{y}})$ is optimal in (3.1).

3.2 Monotonic Separability

A useful sufficient condition for regional decomposability is what we might call *monotonic separability* of $W(\boldsymbol{u})$. It obtains when for any partition $\boldsymbol{u} = (\boldsymbol{u}^1, \boldsymbol{u}^2)$, $W(\bar{\boldsymbol{u}}^1) \geqslant W(\boldsymbol{u}^1)$ and $W(\bar{\boldsymbol{u}}^2) \geqslant W(\boldsymbol{u}^2)$ imply $W(\bar{\boldsymbol{u}}) \geqslant W(\boldsymbol{u})$. In particular, a separable function[1] $W(\boldsymbol{u}) = \sum_i W_i(u_i)$ is monotonically separable. Then we have

Proposition 3.2 *If $W(\boldsymbol{u})$ is monotonically separable, problem* (3.1) *is regionally decomposable.*

Proof Let $(\bar{y}, \bar{\boldsymbol{u}})$ be optimal in (3.1) and $\hat{\boldsymbol{u}}^k$ optimal in (3.4) for $k = 1, \ldots, r$. We wish to show that $(\hat{\boldsymbol{u}}^1, \ldots, \hat{\boldsymbol{u}}^r, \bar{y})$ is optimal in (3.1). We first note that $(\hat{\boldsymbol{u}}^1, \ldots, \hat{\boldsymbol{u}}^r, \bar{y})$ is feasible in (3.1), by hypothesis. To show that $(\hat{\boldsymbol{u}}^1, \ldots, \hat{\boldsymbol{u}}^r, \bar{y})$ is optimal, it suffices to show

$$W(\hat{\boldsymbol{u}}^1, \ldots, \hat{\boldsymbol{u}}^r) \geqslant W(\boldsymbol{u}^1, \ldots, \boldsymbol{u}^r) \tag{3.5}$$

for any $(\boldsymbol{u}^1, \ldots, \boldsymbol{u}^r, \boldsymbol{y})$ that is feasible in (3.1). But if $(\boldsymbol{u}^1, \ldots, \boldsymbol{u}^r, \boldsymbol{y})$ is feasible in (3.1), then $\boldsymbol{u}^k$ is feasible in (3.4) with $\bar{y}_k = y_k$ for each k. Since $\hat{\boldsymbol{u}}^k$ is optimal in (3.4), we have $W(\hat{\boldsymbol{u}}^k) \geqslant W(\boldsymbol{u}^k)$ for each k. Now (3.5) follows from the fact that $W(\boldsymbol{u})$ is monotonically separable. □

Thus if the SWF is monotonically separable, the regions will distribute their allotment in a way consistent with a nationally optimal solution. However, if the SWF is not monotonically separable, or some regions use a different social welfare criterion, a region's distribution to its subregions may depart from the national plan. If so, the nation's allocation of resources to regions cannot be based on the assumption that the resources will be distributed within regions as prescribed by the national solution.

In a network with more than two levels, the definition of collapsibility and Proposition 3.1 essentially remain the same. For regional decomposability, one needs to generalize the definition in a natural way to reflect the consistency of resource allocations at *all* levels. Proposition 3.2 remains the same.

References

1. Simchi-Levi, D., Trichakis, N., Zhang, P.Y.: Designing response supply chain against bioattacks. Operations Research **67**(5), 1246–1268 (2019)
2. Dalton, H.: The measurement of the inequality of incomes. Economic Journal **30**, 348–461 (1920)
3. Atkinson, A.B.: On the measurement of inequality. Journal of Economic Theory **2**, 244–263 (1970)

[1] A common assumption made by early research in this area, such as [2, 3]. For more discussion on such a welfare function, we refer readers to Sect. 8.

Incentives and Sharing

4

4.1 Role of Marginal Analysis

Marginal analysis can inform the choice of a social welfare function by addressing two important and related questions. One is the question of *incentives*: does an SWF incentivize a stakeholder to improve efficiency? This occurs when a stakeholder i can receive a larger utility allotment by using resources more efficiently, that is, by reducing the cost coefficient a_i in the resource constraint. Another is the question of *sharing*: is some of the additional utility generated by increased efficiency shared with others? If so, then the SWF incentivizes *cooperation*. If the reverse occurs, so that efficiency improvements are rewarded in part by transferring utility from others, then the SWF incentivizes *competition*.

Many factors contribute to stakeholder conversion efficiencies. For purposes of our discussion, we categorize them into personal and technological factors. Personal factors are those that pertain to the individual stakeholder, and may or may not be under the stakeholder's control. In the healthcare domain, for example, a patient's choice of diet and exercise regimen can influence the effectiveness of treatment. The type of illness afflicting the patient is a personal factor beyond the patient's control. Technological factors are those that affect the efficiency of stakeholders in general, and can sometimes be influenced by social and industrial policy. It may be possible, for example, to develop effective and affordable drugs for certain illnesses.

When personal factors are under the stakeholder's control, it is obviously important to select an SWF that incentivizes individual efficiency improvements. The sharing properties of an SWF can also influence efficiency if stakeholders respond differently in a cooperative or competitive environment. Sharing properties are also obviously important when planners place a value on incentivizing cooperation or competition for its own sake.

Ö. Elçi et al., *The Structure of Fair Solutions*, Synthesis Lectures on Operations Research and Applications, https://doi.org/10.1007/978-3-031-82190-5_4

Technological factors can also help influence the choice of an SWF, because different SWFs incentivize investment in different technologies. Some SWFs recognize a greater improvement in social welfare when technology boosts the efficiency of less efficient stakeholders, while others favor technologies that benefit stakeholders that are already more efficient. The choice of SWF can therefore lead to less or greater inequality by encouraging investment in technologies that have the corresponding effect.

4.2 Formulas for Marginal Analysis

We carry out marginal analysis by observing how reducing a stakeholder i's cost coefficient a_i in the budget constraint changes the socially optimal utility distribution $\boldsymbol{u}^*$. For purposes of this analysis, we assume the optimum is obtained subject only to a resource constraint, without upper bounds on utilities. We focus on effects at the margin, which means that we investigate only the effect of small perturbations in a_i. To simplify the analysis without undermining basic insights, we ignore points at which the solution $\boldsymbol{u}^*$ is a nonsmooth function of $\boldsymbol{a}$. This allows us to express the percentage increase in a stakeholder's utility u_j^* for each percent decrease in a_i in terms of partial derivatives:

$$-\frac{\partial u_j^*/u_j^*}{\partial a_i/a_i} = -\frac{\partial u_j^*}{\partial a_i} \cdot \frac{a_i}{u_j^*} \tag{4.1}$$

We refer to (4.1) as the *incentive rate* for stakeholder i when $j = i$ and as the *cross-subsidy rate* from stakeholder i to j when $j \neq i$. The incentive rate indicates stakeholder i's incentive to upgrade conversion efficiency. The cross-subsidy rates indicate the degree to which the fruits of an upgrade are shared with other stakeholders. One can interpret negative cross-subsidy rates as encouraging competition, and positive rates as encouraging cooperation.

We also examine implications for social planning in the form of technology investments in the improvement of conversion efficiencies. The rate of social welfare improvement that results from percentage increase in stakeholder i's efficiency is

$$\delta_i = -\frac{\partial W(\boldsymbol{u}^*)}{\partial a_i/a_i} = -\frac{\partial W(\boldsymbol{u}^*)}{\partial a_i} \cdot a_i$$

This results in a gradient $(\delta_1, \ldots, \delta_n)$, which we refer to as the *technology payoff gradient*. It indicates the direction of the steepest ascent, that is, the relative percentage improvements in individual stakeholder efficiencies that would result in the largest rate of increase in social welfare. Thus, to maximize the social welfare growth rate at the margin, stakeholders $1, \ldots, n$ should achieve percentage growth rates in efficiency that are proportional to $\delta_1, \ldots, \delta_n$, respectively.

Utilitarian and Maximin Criteria 5

5.1 Mathematical Formulation

An analysis of utilitarian and maximin social welfare functions is straightforward but a useful starting point. The *utilitarian* SWF

$$W(\boldsymbol{u}) = \sum_i u_i$$

is most characteristic of traditional objective functions. It stems ultimately from Jeremy Bentham's idea that one should maximize the greatest good for the greatest number [1]. The generic social welfare optimization problem (2.1) becomes simply

$$\max_{\boldsymbol{u},\boldsymbol{x}} \left\{ \sum_i u_i \;\middle|\; (\boldsymbol{u},\boldsymbol{x}) \in S \right\}$$

The *maximin* criterion is inspired by the Difference Principle of John Rawls [2], which has been discussed in a large literature [3, 4]. It states roughly that inequality is justified only to the extent that it improves the lot of the worst off. The criterion, therefore, seeks to maximize the minimum utility and corresponds to the simple SWF

$$W(\boldsymbol{u}) = \min_i \{u_i\}$$

While Rawls intended the principle only to apply to the design of social institutions and the distribution of "primary goods," it can be investigated as a possible rule for fair distribution in general. The maximin SWF can be formulated for the generic model (2.1) by maximizing an auxiliary variable v that represents the smallest utility:

$$\max_{\boldsymbol{u},\boldsymbol{x},v} \left\{ v \;\middle|\; v \le u_i,\ \text{all } i;\ (\boldsymbol{u},\boldsymbol{x}) \in S \right\}$$

Ö. Elçi et al., *The Structure of Fair Solutions*, Synthesis Lectures on Operations Research and Applications, https://doi.org/10.1007/978-3-031-82190-5_5

5.2 Socially Optimal Distributions

While most optimization models are designed to maximize total utility (or minimize total cost) in some sense, this can lead to extreme distributions. We can see this by maximizing utility subject to a budget constraint and upper bounds d_i on the utilities. Suppose that utilities are indexed so that $a_1 \leq \cdots \leq a_n$, which means that utility is least expensive for stakeholder 1. Then total utility is maximized by allocating all available utility to stakeholder 1, up to the bound d_1, then all remaining utility to stakeholder 2 up to d_2, and so forth. Thus we have the simple result.

Proposition 5.1 *If* $a_1 \leq \cdots \leq a_n$ *and*

$$\phi_i = \frac{1}{a_i}\Big(B - \sum_{j=1}^{i-1} a_j d_j\Big), \textit{ for } i = 1, \ldots, n$$

then maximizing utility subject to a budget constraint and upper bounds yields the optimal solution:

$$u_i^* = \begin{cases} d_i \textit{ for } i = 1, \ldots, k \\ \phi_i \textit{ for } i = k+1 \\ 0 \;\; \textit{for } i = k+2, \ldots, n \end{cases}$$

where $k = \max\{i \in \{0, \ldots, n\} \mid d_i \leq \phi_i\}$. *In particular, if* $B/a_1 \leq d_1$, *then stakeholder 1 receives all available utility, and everyone else receives nothing.*

Proof Since (2.4) is a linear programming problem, it suffices to exhibit a feasible solution of the dual problem whose objective function value is equal to the optimal value $\sum_i u_i^*$ of the primal. The dual problem is

$$\min\Big\{Bv_0 + \sum_{j \in N} d_i v_i \;\Big|\; v_0 \geq 0;\; a_i v_0 + v_i \geq 1,\; v_i \geq 0 \text{ for } i \in N\Big\}$$

If $u_n^* = d_n$, the desired dual solution is $v_0 = 0$, $v_i = 1$ for $i \in N$. Its objective value is the same as the optimal value $\sum_i d_i$ of (2.4). Otherwise, the desired dual solution is

$$v_i = \begin{cases} B/a_k & \text{for } i = 0 \\ 1 - a_i/a_k & \text{for } i = 1, \ldots, k-1 \\ 0 & \text{for } i = k, \ldots, n \end{cases}$$

where $k = \min\{i \in N \mid u_i^* < d_i\}$. It is easily verified that this solution is dual feasible and has the same objective function value as the optimal solution of (2.4), namely, $B/a_k + \sum_{i=1}^{k-1}(1 - a_i/a_k)d_i$.

While a utilitarian policy of lavishing all resources on a single or a few stakeholders is often unacceptable, a maximin criterion can also lead to extreme outcomes. The maximin solution subject to a budget constraint splits utility equally among stakeholders. Thus if we define $a(i) = \sum_{j=1}^{i} a_j$, we have $u_i^* = B/a(n)$ for all i, so that stakeholder i receives resource allotment $Ba_i/a(n)$. This can require a very unequal distribution of resources when the conversion efficiencies differ, as illustrated in the second line of Table 2.2. A more extreme case arises when devoting all available resources to a seriously ill patient yields only a slight improvement. The maximin criterion devotes nearly all resources to that person and reduces the welfare of everyone else to the same low level. The outcome is similar if there are upper bounds. If we let $d_{\min} = \min_i\{d_i\}$, then we have the following result.

Proposition 5.2 *A maximin solution subject to a budget constraint and upper bounds allots equal utility* $u_i^* = \min\{d_{\min}, B/a(n)\}$ *to all stakeholders* i.

Proof Let $d_k = d_{\min}$. If the given solution $\boldsymbol{u}^*$ is not maximin, there must be a feasible solution $\bar{\boldsymbol{u}}$ such that $\bar{u}_i > u_i^*$ for all i. In particular, since $\bar{u}_k > u_k^* = \min\{d_k, B/a(n)\}$ and $\bar{u}_k \leq d_k$, we must have $B/a(n) < d_k$. Now since each $a_i > 0$,

$$\sum_i a_i \bar{u}_i > \sum_i a_i u_i^* = a(n) \min\left\{d_k, \frac{B}{a(n)}\right\} = \frac{Ba(n)}{a(n)} = B$$

which implies that $\bar{\boldsymbol{u}}$ is infeasible.

The proposition reveals another problem with the maximin criterion. A solution in which the utilities are equal to $d_{\min}$ can waste much of the available resources, specifically $\max\{0, B - a(n)d_{\min}\}$. This anomaly is avoided by moving to a leximax criterion, which is presented in the next chapter.

5.3 Hierarchical Distributions

The hierarchical problem (3.1) is regionally decomposable for the utilitarian criterion because the SWF is separable and therefore monotonically separable. The maximin SWF is not separable, but it is monotonically separable and therefore regionally decomposable due to Proposition 3.2.

The utilitarian criterion is collapsible only in the degenerate case where $B = B_k = 0$ for all $k \neq 1$. The utilitarian solution is again extreme. If there are no upper bounds, for example, at most one subregion in each region receives positive utility.

A straightforward argument shows that the maximin problem is collapsible when

$$B + \sum_k B_k \geq \min\left\{a(n)d_{\min},\ \max_k \left\{\frac{a(n)}{a(I_k)} B_k\right\}\right\}$$

where $d_{\min} = \min_i\{d_i\}$ and $a(I_k) = \sum_{i \in I_k} a_i$. An optimal solution of the collapsed problem is

$$u_i^* = \min\left\{d_{\min}, \frac{1}{a(n)}\Big(B + \sum_k B_k\Big)\right\}, \text{ all } i$$

5.4 Incentives and Sharing

Finally, we examine the effect on a socially optimal solution $\boldsymbol{u}^*$ (subject to a budget constraint) when a stakeholder i improves personal efficiency by reducing a_i. In the utilitarian problem, only the stakeholder who receives all resources ($i = 1$) benefits from efficiency improvements at the margin, and everyone else continues to receive zero utility. Thus the incentive rate is 1 for the most efficient stakeholder and 0 for everyone else. The cross-subsidy rates are all 0, and the technology payoff gradient is a positive scalar multiple of $(1, 0, \ldots, 0)$, meaning that, at the margin, society should invest only in the most efficient stakeholder.

In the maximin problem, the incentive and cross-subsidy rates for stakeholder i are

$$\frac{\partial u_j^*}{\partial a_i} \cdot \frac{a_i}{u_j^*} = \frac{a_i}{a(n)}, \quad \text{all } j$$

This means that when stakeholder i improves efficiency, the surplus is shared with other stakeholders, all of whom receive the same percentage of their current cost in benefits. The incentive to improve is roughly proportional to the current cost a_i. Those who waste more resources have a stronger incentive to cut waste. The technology payoff gradient is a positive scalar multiple of $(a_1, \ldots, a_n)$, indicating that society should direct greater investment to less efficient stakeholders.

References

1. Bentham, J.: An Introduction to the Principles of Morals and Legislation. reprinted by Oxford University Press, 1907 (1789)
2. Rawls, J.: A Theory of Justice (revised). Harvard University Press, original edition 1971 (1999)
3. Richardson, H.S., Weithman, P.J. (eds.): The Philosophy of Rawls (5 volumes). Garland (1999)
4. Freeman, S. (ed.): The Cambridge Companion to Rawls. Cambridge University Press (2003)

The Leximax Criterion 6

6.1 Mathematical Formulation

A *leximax* (lexicographic maximization) criterion is an extension of the maximin criterion.[1] It maximizes the smallest utility, then maximizes the second smallest utility while holding the smallest fixed, and so forth until all utilities are determined in this way. In so doing it avoids at least two weaknesses of the maximin criterion. It necessarily makes use of all available resources (unless utility bounds prevent this), and it accounts for the equity situation of stakeholders other than the very worst off. In addition, the Rawlsian defense of a maximin criterion can be plausibly extended to a leximax criterion. However, identifying a leximax distribution can pose a challenging computational problem when there are many stakeholders.

Lexicographic maximization is defined differently in a fairness context than in most other contexts. When comparing utility distributions $\boldsymbol{u}$, it assumes that the utilities in $\boldsymbol{u}$ are ordered by size rather than fixed positions in a sequence $u_1, \ldots, u_n$. To make this precise, we say that utility distribution $\boldsymbol{u}$ *dominates* $\boldsymbol{u}'$ when there is a permutation $\boldsymbol{\pi}$ of $1, \ldots, n$ for which $u_{\pi_1} \leq \cdots \leq u_{\pi_n}$, a permutation $\boldsymbol{\pi}'$ for which $u'_{\pi'_1} \leq \cdots \leq u'_{\pi'_n}$, and an index $k \in N$, such that the following holds: $u_{\pi_j} = u'_{\pi'_j}$ for $j = 1, \ldots, k-1$ and $u_{\pi_k} > u'_{\pi'_k}$. Thus, while $\boldsymbol{u} = (1, 3, 2)$ dominates $\boldsymbol{u}' = (1, 2, 4)$ in the usual sense, $\boldsymbol{u}'$ dominates $\boldsymbol{u}$ in the sense defined here. We say that $\boldsymbol{u}^* \in U$ is a leximax solution over a feasible set U if no element of U dominates $\boldsymbol{u}^*$.

[1] Confusingly, a leximax criterion is sometimes referred to as leximin, and sometimes as lexmaxmin. We use the term "leximax" because it seems more prevalent in the optimization literature.

Ö. Elçi et al., *The Structure of Fair Solutions*, Synthesis Lectures on Operations Research and Applications, https://doi.org/10.1007/978-3-031-82190-5_6

The leximax problem can be formulated as a sequence of n optimization problems, provided that one is careful to examine all alternate optimal solutions. Problem k in the sequence determines the kth smallest utility of the leximax solution:

$$\max_{\boldsymbol{u},\boldsymbol{x},\pi,v} \left\{ v \;\middle|\; \begin{array}{l} u_{\pi_j} = u^*_{\pi_j},\ j = 1, \ldots, k-1 \\ v \le u_i,\ i \in \{1, \ldots, n\} \setminus \{\pi_1, \ldots, \pi_{k-1}\} \\ (\boldsymbol{u}, \boldsymbol{x}) \in S \end{array} \right\} \tag{6.1}$$

Here, $u^*_{\pi_j}$ is an optimal value obtained for u_{π_j} in a previous problem j. Each π_j is chosen so that $u^*_{\pi_j}$ is a smallest utility obtained in problem j, that is,

$$u^*_{\pi_j} = \min\left\{\hat{u}_i \;\middle|\; i \in \{1, \ldots, n\} \setminus \{\pi_1, \ldots, \pi_{j-1}\}\right\}$$

where $\hat{u}_1, \ldots, \hat{u}_n$ are the optimal values obtained in problem j. There may be several possible choices for π_j after solving problem j, because there may be a many-way tie for the smallest utility. Each problem j must be solved for all possible sequences of $\pi_1, \ldots, \pi_{j-1}$ inherited from previous problems. The alternate solutions $(u^*_1, \ldots, u^*_n)$ obtained at the end of this process can then be examined to identify one that is undominated by the others and is therefore a leximax solution.

As an example, suppose that (u_1, u_2, u_3) has two feasible values, $(1, 3, 2)$ and $(1, 2, 4)$. Then problem $k = 1$ is

$$\max_{\boldsymbol{u},v} \left\{v \;\middle|\; v \le u_1,\ v \le u_2,\ v \le u_3,\ \boldsymbol{u} \in \{(1, 3, 2), (1, 2, 4)\}\right\}$$

Both $(1, 3, 2)$ and $(1, 2, 4)$ are optimal solutions, and in either case $u^*_{\pi_1} = u^*_1 = 1$. Thus only one value of π_1 need be considered, namely, $\pi_1 = 1$. Problem $k = 2$ is

$$\max_{\boldsymbol{u},v} \left\{v \;\middle|\; u_1 = 1,\ v \le u_2,\ v \le u_3,\ \boldsymbol{u} \in \{(1, 3, 2), (1, 2, 4)\}\right\}$$

One optimal solution is $\hat{\boldsymbol{u}} = (1, 3, 2)$, with $u^*_{\pi_2} = u^*_3 = 2$. Another optimal solution is $\hat{\boldsymbol{u}} = (1, 2, 4)$, with $u^*_{\pi_2} = u^*_2 = 2$. Thus π_2 can be 3 or 2. If we select $\pi_2 = 3$, problem $k = 3$ is

$$\max_{\boldsymbol{u},v} \left\{v \;\middle|\; (u_1, u_3) = (1, 2),\ v \le u_2,\ \boldsymbol{u} \in \{(1, 3, 2), (1, 2, 4)\}\right\}$$

with optimal solution $\hat{\boldsymbol{u}} = (1, 3, 2)$, which sets $u^*_{\pi_3} = u^*_2 = 3$. If we select $\pi_2 = 2$, problem $k = 3$ is

$$\max_{\boldsymbol{u},v} \left\{v \;\middle|\; (u_1, u_2) = (1, 2),\ v \le u_3,\ \boldsymbol{u} \in \{(1, 3, 2), (1, 2, 4)\}\right\}$$

with optimal solution $\hat{\boldsymbol{u}} = (1, 2, 4)$, which sets $u^*_{\pi_3} = u^*_3 = 4$. The first solution yields $\boldsymbol{u}^* = (1, 3, 2)$ and the second $\boldsymbol{u}^* = (1, 2, 4)$. Only the former is a leximax solution, as can be checked by noting which solution dominates the other. It is therefore necessary to examine both possible values of (π_1, π_2, π_3) to be sure of finding a leximax solution.

Ogryczak and Śliwiński [1] show how to obtain a leximax solution with a single optimization model. However, the model may require very large coefficients in the objective function, which may create numerical problems due to roundoff error.

6.2 Socially Optimal Distribution

A leximax distribution, subject to a budget constraint and upper bounds on utilities, can be obtained as follows. Set the utilities with the k smallest upper bounds equal to these upper bounds, and then set the remaining utilities to a common value that is selected to exhaust the remaining resources. If there are no upper bounds, the leximax solution is the same as the maximin solution. The number k depends on the specific problem data, as described in the following theorem:

Proposition 6.1 *Suppose that utilities* $u_1, \ldots, u_n$ *are indexed so that their upper bounds satisfy* $d_1 \leq \cdots \leq d_n$*, and that*

$$\psi_i = \frac{1}{a(i:n)}\Big(B - \sum_{j=1}^{i-1} a_j d_j\Big), \quad \textit{for } i = 1, \ldots, n$$

Then a leximax solution $\boldsymbol{u}^*$ *subject to a budget constraint and the upper bounds is given by* $u_i^* = d_i$ *for all* i *if* $\psi_n \geq d_n$*, and otherwise by*

$$u_i^* = \begin{cases} d_i & \textit{for } i = 1, \ldots, k-1 \\ \psi_k & \textit{for } i = k, \ldots, n \end{cases} \tag{6.2}$$

where $k = \min\{i \mid \psi_i < d_i\}$*. Furthermore, solution* (6.2) *consumes all available resources. If there are no upper bounds, then the maximin solution* $u_i^* = B/a(n)$ *for all* i *is a leximax solution.*

Proof We first suppose that $\psi_n \geq d_n$. From the definition of ψ_n, this implies that $\sum_i a_i d_i \leq B$, and a leximax solution must therefore fix each u_i^* to its upper bound d_i. We next suppose that $\psi_n < d_n$ and show first that the solution (6.2) consumes all available resources B. Resource consumption is

$$\sum_{i=1}^{n} a_i u_i^* = \sum_{i=1}^{k-1} a_i d_i + \sum_{i=k}^{n} a_i \psi_k \tag{6.3}$$

Substituting the definition of ψ_k, we find that (6.3) simplifies to B. We also note that by definition of k and the fact that $d_k \leq \cdots \leq d_n$, we have $u_i^* = \psi_k < d_i$ for $i = k, \ldots, n$. Thus all and only the $k-1$ smallest utilities in $\boldsymbol{u}^*$ are set to their upper bounds. We now suppose, contrary to the claim, that $\boldsymbol{u}^*$ is dominated by some feasible solution $\boldsymbol{u}'$. Since the $k-1$ smallest utilities in $\boldsymbol{u}^*$ are set to their upper bounds, the $k-1$ smallest utilities in $\boldsymbol{u}'$ cannot exceed the $k-1$ smallest utilities in $\boldsymbol{u}^*$, respectively, and must therefore be equal to these utilities. Thus if $\boldsymbol{u}'$ is to dominate $\boldsymbol{u}^*$, some utility among the $n-k+1$ largest in $\boldsymbol{u}'$ must exceed the common value ψ_k of $u_k^*, \ldots, u_n^*$. But this forces another of the $n-k+1$ largest utilities in $\boldsymbol{u}'$ to be smaller than ψ_k, because the utilities $\boldsymbol{u}^*$ consume all available resources. This is inconsistent with the assumption that $\boldsymbol{u}'$ dominates $\boldsymbol{u}^*$. Finally, if there are no upper bounds, we can set each $d_i = \infty$ and note that $k = 1$. Thus each $u_i^* = \psi_1 = B/a(n)$.

As an example, suppose that we desire a leximax distribution subject to the budget constraint $10u_1 + 5u_2 + 3u_3 + 2u_4 \le 100$ and bounds $(d_1, \ldots, d_4) = (3, 6, 10, 12)$. Then $(\psi_1, \ldots, \psi_4) = (5, 7, 8, 5)$ and $k = 3$. This yields the leximax solution $\boldsymbol{u}^* = (3, 6, 8, 8)$. If there are no upper bounds, the maximin solution $\boldsymbol{u}^* = (5, 5, 5, 5)$ is a leximax solution.

Since the leximax solution without upper bounds is the same as the maximin solution, it is likewise capable of delivering extreme solutions in which some stakeholders must make great sacrifices to achieve equality with less efficient stakeholders.

6.3 Hierarchical Distributions

Because the leximax criterion is not represented by a single SWF, we cannot establish regional decomposability by showing monotone separability (Proposition 3.2). However, a property analogous to monotone separability can be demonstrated as follows for lexicographic comparisons, and it will allow us to prove regional decomposability. We write $\bar{\boldsymbol{u}} \succcurlyeq \boldsymbol{u}$ when $\bar{u}_{\langle j \rangle} \ge u_{\langle j \rangle}$ for $j = 1, \ldots, n$, where $(\bar{u}_{\langle 1 \rangle}, \ldots, \bar{u}_{\langle n \rangle})$ is $\bar{\boldsymbol{u}}$ arranged in non-decreasing order, and similarly for $(u_{\langle 1 \rangle}, \ldots, u_{\langle n \rangle})$. Then the desired property is given in the following lemma, whose proof is based on a bipartite matching:

Lemma 6.1 *If $\bar{\boldsymbol{u}} \succcurlyeq \boldsymbol{u}$ and $\bar{\boldsymbol{v}} \succcurlyeq \boldsymbol{v}$, then $(\bar{\boldsymbol{u}}, \bar{\boldsymbol{v}}) \succcurlyeq (\boldsymbol{u}, \boldsymbol{v})$.*

Proof Let $(\bar{w}_1, \ldots, \bar{w}_n)$ consist of the components of $(\bar{\boldsymbol{u}}, \bar{\boldsymbol{v}})$ in non-decreasing order, and similarly for $(w_1, \ldots, w_n)$. It suffices to show that $\bar{w}_k \ge w_k$ for $k = 1, \ldots, n$. We first note that there is a perfect matching between $\bar{w}_1, \ldots, \bar{w}_n$ and $w_1, \ldots, w_n$ in which each $\bar{u}_{\langle j \rangle}$ is matched with $u_{\langle j \rangle}$ and each $\bar{v}_{\langle j \rangle}$ with $v_{\langle j \rangle}$. For any given k, $\bar{w}_k$ must be matched with w_ℓ either for $\ell \ge k$ or for $\ell < k$. If $\ell \ge k$, then $\bar{w}_k \ge w_\ell \ge w_k$, so that $\bar{w}_k \ge w_k$. If $\ell < k$, then $\bar{w}_i$ for some $i < k$ must be matched with w_j for some $j \ge k$. Now we have $\bar{w}_k \ge \bar{w}_i \ge w_j \ge w_k$, which again implies $\bar{w}_k \ge w_k$.

Regional decomposability is defined for the leximax problem by replacing $\max\{W(\boldsymbol{u})\}$ with $\text{leximax}\{\boldsymbol{u}\}$ in (3.1) and (3.4). We now have the following result:

Proposition 6.2 *The hierarchical problem (3.1) is regionally decomposable for a leximax criterion.*

Proof Let $(\bar{y}, \bar{\boldsymbol{u}})$ be a leximax solution of (3.1), which means that $\bar{\boldsymbol{u}} \succcurlyeq \boldsymbol{u}$ for all $(\boldsymbol{u}, \boldsymbol{y})$ that are feasible in (3.1). Let $\hat{\boldsymbol{u}}^k$ a leximax solution of (3.4) for each k, so that

$$\hat{\boldsymbol{u}}^k \succcurlyeq \boldsymbol{u}^k,\ k = 1, \ldots, r \tag{6.4}$$

for all $\boldsymbol{u}^k$ feasible in (3.4). We note that by construction, $(\hat{\boldsymbol{u}}^1, \ldots, \hat{\boldsymbol{u}}^r, \bar{y})$ is feasible in (3.1). We wish to show that

$$(\hat{\boldsymbol{u}}^1, \ldots, \hat{\boldsymbol{u}^r}) \succcurlyeq (\boldsymbol{u}^1, \ldots, \boldsymbol{u}^r) \tag{6.5}$$

for any $(\boldsymbol{u}^1, \ldots, \boldsymbol{u}^r, \boldsymbol{y})$ feasible in (3.1). But for any such solution of (3.1), $\boldsymbol{u}^k$ is feasible in (3.4) with $\bar{y}_k = y_k$. Thus (6.4) holds, which implies (6.5) by Lemma 6.1.

6.4 Incentives and Sharing

Since the leximax solution subject to a budget constraint is the same as the maximin solution, the analysis of incentives and cross subsidies is the same (Section 5.4).

Reference

1. Ogryczak, W., Sliwinski, T.: On direct methods for lexicographic min-max optimization. In: M. Gavrilova, O. Gervasi, V. Kumar, C.J.K. Tan, D. Taniar, A. Laganá, Y. Mun, H. Choo (eds.) Proceedings of International Conference on Computational Science and Its Applications (ICCSA 2006), *LNCS*, vol. 3982, pp. 802–811 (2006)

Beta Fairness 7

7.1 Mathematical Formulation

Beta fairness generalizes the maximin criterion, as does the leximax criterion, but with less computational complexity. It maximizes the average welfare of a disadvantaged class of stakeholders at the bottom of the distribution, rather than the welfare of a single stakeholder. As it turns out, however, beta-fair solutions tend to be simple maximin solutions except when most of the population is viewed as disadvantaged.

Specifically, beta fairness maximizes the *conditional β-mean*, which is the average utility of the least advantaged $100\beta\%$ of the population [1, 2]. We will call this the *reference group*. Thus the SWF is

$$W(\boldsymbol{u}) = \frac{1}{\beta n} \sum_{i=1}^{\beta n} u_{\langle i \rangle}$$

where $u_{\langle 1 \rangle}, \ldots, u_{\langle n \rangle}$ are $u_1, \ldots, u_n$ arranged in nondecreasing order, breaking ties arbitrarily. We assume βn is a positive integer k, where k is the size of the reference group. Beta fairness is a maximin criterion when $k = 1$ and a purely utilitarian criterion when $k = n$. It is interesting to see how it compromises between the two when $1 < k < n$.

Beta fairness might initially be seen as preventing a relatively small group at the bottom of the distribution from being overly disadvantaged. This would call for a relatively small β in applications, closer to 0 than 1. We will find, however, that achieving beta fairness (subject to a linear budget constraint) generally results in a maximin solution in such cases. Beta fairness may therefore be more useful for preventing a relatively small group at the top from being overly privileged, which suggests that β should be closer to 1 than 0.

The problem of maximizing the beta fairness SWF can be given a mathematical programming formulation as follows [2]. For a given vector $\boldsymbol{u}$ of utilities, the average of the $k = \beta n$ smallest utilities (known as the *conditional β-mean* [3]) is

Ö. Elçi et al., *The Structure of Fair Solutions*, Synthesis Lectures on Operations Research and Applications, https://doi.org/10.1007/978-3-031-82190-5_7

$$\min_{\boldsymbol{\delta}} \left\{ \frac{1}{k} \sum_i u_i \delta_i \;\middle|\; \sum_i \delta_i = k,\ \delta_i \in \{0,1\},\ \text{all } i \right\} \tag{7.1}$$

Since the constraint matrix in (7.1) is totally unimodular, and k is integral, the integrality restriction on $\boldsymbol{\delta}$ can be dropped:

$$\min_{\boldsymbol{\delta}} \left\{ \frac{1}{k} \sum_i u_i \delta_i \;\middle|\; \sum_i \delta_i = k,\ 0 \le \delta_i \le 1,\ \text{all } i \right\} \tag{7.2}$$

Since (7.2) is a linear programming problem, we can solve it for a given $\boldsymbol{u}$ by solving its dual:

$$\max_{\lambda,\boldsymbol{\mu}} \left\{ k\lambda - \sum_i \mu_i \;\middle|\; \lambda - \mu_i \le \frac{u_i}{k},\ \mu_i \ge 0,\ \text{all } i \right\} \tag{7.3}$$

We can now achieve beta fairness by maximizing (7.3) over all feasible $\boldsymbol{u}$, that is, all $\boldsymbol{u}$ such that $(\boldsymbol{u}, \boldsymbol{x}) \in S$ for some $\boldsymbol{x}$. This can be accomplished by treating $\boldsymbol{u}$ as a variable in (7.3) that is constrained by $(\boldsymbol{u}, \boldsymbol{x}) \in S$, and we have the following:

Proposition 7.1 *A beta-fair distribution $\boldsymbol{u}$ subject to $(\boldsymbol{u}, \boldsymbol{x}) \in S$ can be obtained by solving*

$$\max_{\boldsymbol{u},\boldsymbol{x},\lambda,\boldsymbol{\mu}} \left\{ k\lambda - \sum_i \mu_i \;\middle|\; \lambda - \mu_i \le \frac{u_i}{k},\ \mu_i \ge 0,\ \text{all } i;\ (\boldsymbol{u}, \boldsymbol{x}) \in S \right\} \tag{7.4}$$

where $k = \beta n$.

7.2 Socially Optimal Solution

A distribution that achieves beta fairness subject to a budget constraint has a distinct structure that depends on β and the relative size of the coefficients $a_1, \ldots, a_n$, but not on the size of the budget B. Again letting $k = \beta n \ge 1$ be the size of the reference group, utility is equally divided among the ℓ most efficient stakeholders when

$$\frac{a(\ell)}{a_{\ell+1}} + n - \ell \le k \le \frac{a(\ell)}{a_\ell} + n - \ell \tag{7.5}$$

if we define $a_{n+1} = \infty$. For example, if $(a_1, \ldots, a_4) = (4, 6, 8, 9)$ and $B = 108$, then we have the maximin solution $\boldsymbol{u} = (4, 4, 4, 4)$ when $k = 1$, the same solution when $k = 2$, the solution $\boldsymbol{u} = (6, 6, 6, 0)$ when $k = 3$, and the utilitarian solution $\boldsymbol{u} = (13, 0, 0, 0)$ when $k = 4$. The general result is established as follows, where as usual $a(\ell) = \sum_{i=1}^{\ell} a_i$.

Proposition 7.2 *If $a_1 \le \cdots \le a_n$, a utility distribution $\boldsymbol{u}$ that achieves beta fairness for subject to a resource constraint is given by*

$$u_i = \begin{cases} \dfrac{B}{a(\ell)}, & \textit{for } i = 1, \ldots, \ell \\ 0, & \textit{for } i = \ell + 1, \ldots, n \end{cases} \tag{7.6}$$

when $k = \beta n$ *for some integer* k*, and* ℓ *satisfies*

$$\frac{a(\ell)}{a_{\ell+1}} \leq k + \ell - n \leq \frac{a(\ell)}{a_\ell} \tag{7.7}$$

or equivalently, when ℓ *satisfies* (7.5)*, with the convention that* $a_{n+1} = \infty$.

Proof The optimization problem is (7.4) in which $\boldsymbol{u} \in S$ is the budget constraint $\boldsymbol{a}^{\mathsf{T}}\boldsymbol{u} \leq B$ and $\boldsymbol{u} \geq 0$. This is a linear programming problem whose dual is the following:

$$\min_{z,y} \left\{ Bz \;\middle|\; \sum_i y_i = k;\ z \geq \frac{y_i}{ka_i},\ 0 \leq y_i \leq 1,\ u_i \geq 0,\ \text{all } i \right\} \tag{7.8}$$

where dual variable z corresponds to the budget constraint $\boldsymbol{a}^{\mathsf{T}}\boldsymbol{u} \leq B$, and y_i corresponds to the constraint $\lambda - \mu_i \leq u_i/k$. We will show that the following primal and dual solutions are feasible and yield the same objective function value, which establishes that both are optimal.

$$(u_i, \mu_i) = \begin{cases} (k\lambda, 0), & \text{for } i = 1, \ldots, \ell \\ (0, \lambda), & \text{for } i = \ell + 1, \ldots, n \end{cases} \quad \text{where } \lambda = \frac{B}{ka(\ell)} \tag{7.9}$$

$$y_i = \begin{cases} ka_i z, & \text{for } i = 1, \ldots, \ell \\ 1, & \text{for } i = \ell + 1, \ldots, n \end{cases} \quad \text{where } z = \frac{k + \ell - n}{k} \cdot \frac{1}{a(\ell)} \tag{7.10}$$

Solution (7.9) is clearly feasible in the primal problem (7.4). To show that (7.10) is feasible in the dual problem (7.8), note first that direct substitution of the solution value of each y_i satisfies $\sum_i y_i = k$. Constraint $y_i \leq 1$ obviously holds for $i > \ell$. For $i \leq \ell$, the constraint becomes

$$(k + \ell - n)\frac{a_i}{a(\ell)} \leq 1$$

which follows from the second inequality in (7.7). The dual constraint $z \geq y_i/ka_i$ holds by direct substitution when $i \leq \ell$. When $i > \ell$, the constraint becomes

$$\frac{k + \ell - n}{a(\ell)} \geq \frac{1}{a_i}$$

which follows from the first inequality in (7.7) because $a_{\ell+1} \leq \cdots \leq a_i$. Finally, the primal and dual objective functions have the same optimal value, namely,

$$\frac{(k + \ell - n)B}{ka(\ell)}$$

which proves the optimality of the solutions (7.9)–(7.10). In particular, the solution value for $\boldsymbol{u}$ in (7.9) is optimal, and it is identical to the solution (7.6) given in the proposition.

A factor that may reduce the applicability of beta fairness is that it yields a maximin solution, or a solution very close to maximin, when β is relatively small. In particular, the solution is maximin ($j = n$) whenever $k \le a(n)/a_n$, in which case all parties receive equal utility. Equivalently, the solution is maximin when the largest cost coefficient a_n is at most $1/\beta$ times the average cost coefficient. Thus if the reference group is 10% of the population, the solution is maximin when the largest cost is less than 10 times the average cost. Furthermore, this is only a sufficient condition for a maximin solution, not a necessary one. For example, the solution given in Table 2.2 for $\boldsymbol{a} = (5, 6, \ldots, 12)$ is maximin for *all* $\beta < 1$. Even when the solution is not maximin, we have $\ell \ge a(\ell)/a_{\ell+1} + n - k \ge n - k$, so that the utility is split among no fewer than $n - k$ or 90% of the stakeholders.

Given these facts, beta fairness is not likely to provide a meaningful alternative to a maximin criterion when the reference group is relatively small. However, a large reference group can be useful when the goal is simply to prevent the most efficient stakeholder from receiving everything, as would occur under a utilitarian criterion. For example, if the cost coefficients are $(a_1, \ldots, a_n) = (1, \ldots, n)$, it is optimal to distribute utility evenly among the $\max\{n, 2(n - k)\}$ most efficient stakeholders. Thus while everyone receives equal utility if less than half the population is in the reference group, only 60% receive positive utility if 70% are in the reference group.

7.3 Hierarchical Distribution

Beta fairness is regionally decomposable for $\beta = 1/n$ and $\beta = 1$, because these represent maximin and utilitarian criteria, respectively. However, it is not regionally decomposable in general for intermediate values of β.

For a counterexample, let $\boldsymbol{a} = (1, 2, 3, 7)$ with top-level budget $B = 0$ and $\beta = \frac{1}{2}$, and suppose there are two regions, each containing two stakeholders, with budgets $(B_1, B_2) = (3, 3)$. The hierarchical problem (3.1) is to maximize the conditional $\frac{1}{2}$-mean subject to

$$\begin{aligned} &y_1 + y_2 \le 0, \quad y_1, y_2 \ge 0 \\ &u_1 + 2u_2 \le 3 + y_1, \quad 3u_3 + 7u_4 \le 3 + y_2, \quad u_i \ge 0, \ \text{all } i \end{aligned}$$

Obviously, we must have $y_1 = y_2 = 0$. By applying Proposition 7.1, we find that the unique optimal solution is $\bar{\boldsymbol{u}} = (1, 1, 1, 0)$. Each of the regional problems is a maximin problem because $2\beta = 1$. Thus the region 1 problem is to maximize $\min\{u_1, u_2\}$ subject to $u_1 + 2u_2 \le 3$, which has the unique solution $\boldsymbol{u}^1 = (u_1, u_2) = (1, 1)$. The region 2 problem is to maximize $\min\{u_3, u_4\}$ subject to $3u_3 + 7u_4 \le 3$, which has the unique solution $\boldsymbol{u}^2 = (u_3, u_4) = (0.3, 0.3)$. Since $(\boldsymbol{u}^1, \boldsymbol{u}^2)$ differs from $\bar{\boldsymbol{u}}$, the problem is not regionally decomposable.

7.4 Incentives and Sharing

Let u_j^* be the value of u_j in a beta-fair solution when $\beta n = k$. Then Proposition 7.2 implies that the cross-subsidy rates for stakeholder $i \in \{1, \ldots, \ell\}$ are

$$-\frac{\partial u_j^*}{\partial a_i} \cdot \frac{a_i}{u_j^*} = \frac{a_i}{a(\ell)}$$

provided $i, j \in \{1, \ldots .\ell\}$, where ℓ satisfies (7.7). If $j = i$, then this expression is the incentive rate for stakeholder i. If $i \notin \{1, \ldots, \ell\}$, then stakeholder i receives no utility, and the incentive rate is zero. If either i or j does not belong to $\{1, \ldots, \ell\}$, then stakeholder i and/or stakeholder j receives no utility, and the cross-subsidy rate is zero.

This implies that a stakeholder that receives positive utility has an incentive to reduce its cost that is approximately proportional to its current cost level. Less efficient utility recipients therefore have a greater incentive to become more efficient. Also a utility recipient i shares part of the benefit of its efficiency improvements with other utility recipients. Each of these other stakeholders benefits by an equal percentage of their current utility, where that percentage is roughly proportional to stakeholder i's current cost. Thus a less efficient utility recipient that improves efficiency by, say, 10% provides greater benefits to both itself and all other utility recipients. Stakeholders that receive no utility are not affected at the margin.

The technology payoff vector is proportional to $(a_1, \ldots, a_\ell, 0, \ldots, 0)$. This means that to maximize the rate of social welfare improvement, available funds for efficiency improvements should, at the margin, be invested in current utility recipients in amounts proportional to their current cost. However, Proposition 7.2 implies that sufficiently large investments in other stakeholders will reallocate positive utility to some of them in a beta-fair solution.

Finally, we note from (7.7) that larger values of β tend to result in fewer stakeholders receiving positive utility. This is as one would expect, since $\beta = 1$ corresponds to a purely utilitarian problem, in which only one stakeholder receives positive utility.

References

1. Ogryczak, W., Zawadzki, M.: Conditional median: A parametric solution concept for location problems. Annals of Operations Research **110**, 167–181 (2002)
2. Filippi, C., Guastaroba, G., Speranza, M.: On single-source capacitated facility location with cost and fairness objectives. European Journal of Operational Research **289**, 959–974 (2021)
3. Ogryczak, W., Śliwiński, T.: On equitable approaches to resource allocation problems: The conditional minimax solutions. Journal of Telecommunications and Information Technology **3**, 40–48 (2002)

8 Alpha Fairness

8.1 Mathematical Formulation

Alpha fairness is one of the most popular criteria for balancing equity and efficiency [1–3]. It has the advantage of allowing one to regulate the balance with a parameter α, where larger values of α place a greater emphasis on fairness. In particular, $\alpha = 0$ corresponds to a purely utilitarian criterion and $\alpha \to \infty$ to a maximin criterion. The alpha fairness SWF is

$$W_\alpha(\boldsymbol{u}) = \begin{cases} \dfrac{1}{1-\alpha} \sum_i u_i^{1-\alpha}, & \text{if } \alpha \geq 0 \text{ and } \alpha \neq 1 \\ \sum_i \log(u_i), & \text{if } \alpha = 1 \end{cases} \tag{8.1}$$

The special case corresponding to $\alpha = 1$ is *proportional fairness*, also known as the *Nash bargaining solution* [4].

Alpha fairness is formulated in an optimization problem (2.1) simply by using one of the two expressions above (depending on the chosen value of α) as the objective function. Contours of the alpha fairness SWF for two stakeholders are illustrated in Fig. 8.1 when $\alpha = 0.5$. The black dot indicates the alpha fair solution if the feasible set is the portion of the nonnegative quadrant under the heavy curve. Stakeholder 1 receives a smaller utility allocation due to the greater difficulty of creating utility for that stakeholder. Contours for $\alpha = 2.0$ appear in Fig. 8.2. Due to a stronger emphasis on fairness, the socially optimal solution allocates somewhat more utility to stakeholder 1.

The solution for proportional fairness ($\alpha = 1$) has an interesting geometric interpretation. It is obtained by selecting a distribution $\boldsymbol{u}^*$ that maximizes the volume $\prod_i u_i$ of a hyperrectangle with opposite corners at the origin and $\boldsymbol{u}$, subject to constraints on $\boldsymbol{u}$. The idea is illustrated for two stakeholders in Fig. 8.3, where the feasible set is the portion of the

Ö. Elçi et al., *The Structure of Fair Solutions*, Synthesis Lectures on Operations Research and Applications, https://doi.org/10.1007/978-3-031-82190-5_8

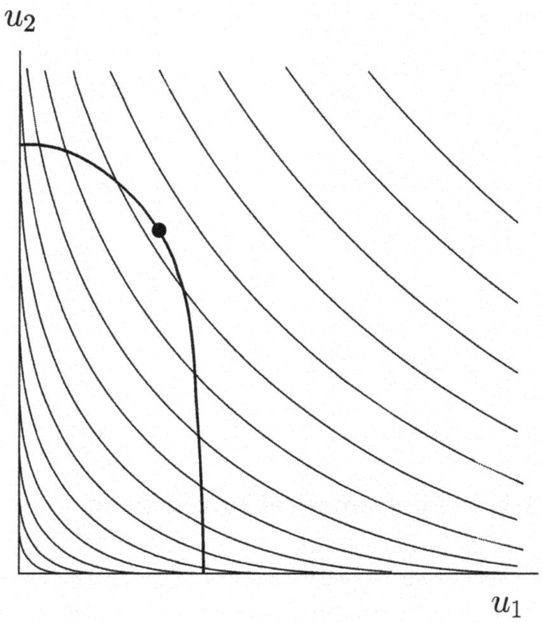

Fig. 8.1 Contours of the 2-person alpha fairness SWF when $\alpha = 0.5$

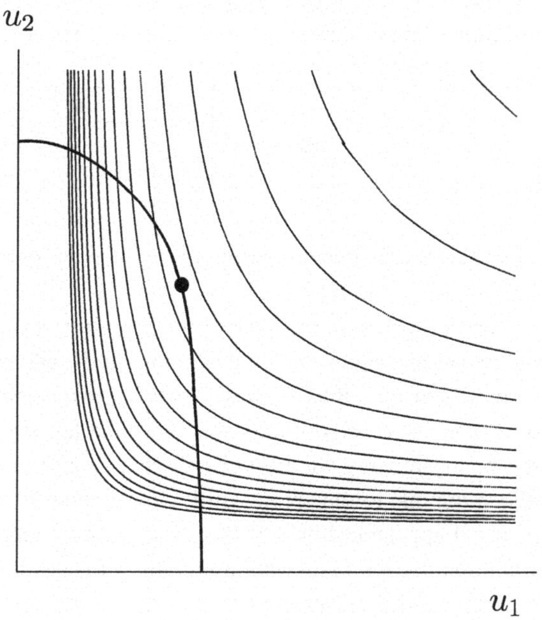

Fig. 8.2 Contours of the 2-person alpha fairness SWF when $\alpha = 2.0$

Fig. 8.3 A Nash bargaining solution for two stakeholders

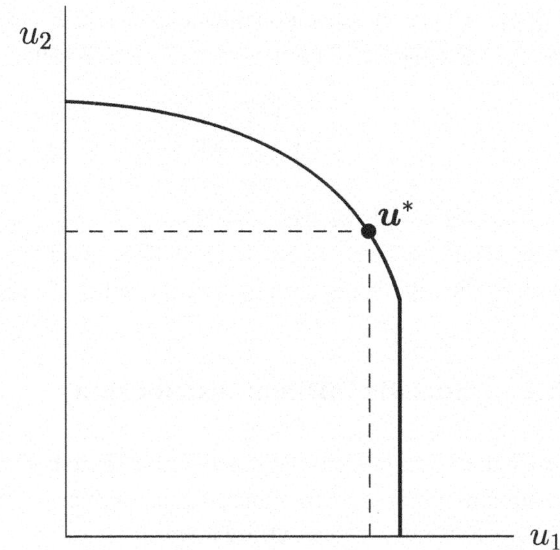

nonnegative quadrant under the heavy curve. The optimal solution $\boldsymbol{u}^*$ maximizes the area of the rectangle defined by the dashed lines.

The alpha fairness criterion has been defended with various axiomatic and bargaining arguments, particularly in the special case of proportional fairness. Nash [4] gave an axiomatic argument for his bargaining solution in 1950, while Rubinstein, Harsanyi, and Binmore [5–7] supplied bargaining arguments. Lan et. al [2, 8, 9] provided an axiomatic derivation for general alpha fairness and proposed an interpretation of the α parameter. Bertsimas et al. [3] studied resulting equity/efficiency trade-offs. Alpha fairness has also seen a number of practical applications, particularly in telecommunications and other engineering fields [1, 10–13].

An issue with alpha fairness is the difficulty in practice of interpreting the parameter α. It is somewhat helpful to characterize mathematically a welfare-preserving transfer of utility from one stakeholder to another. If $u_k > u_j$, then stakeholder k's utility must be reduced by $(u_k/u_j)^\alpha$ to compensate for a one-unit increase in stakeholder j's utility if total social welfare is to remain constant. Thus equality is a stronger imperative for larger α, but it is not obvious what particular value of α is appropriate in a given context. The discussion of incentives and sharing at the end of this chapter (Sect. 8.4) suggests another approach to interpreting α.

Alpha fairness is also capable of extreme solutions in a nonconvex feasible set because when $\alpha \geq 1$, it can assign equality the same social welfare as arbitrarily extreme inequality. This can be seen in Fig. 8.2 where each contour extends infinitely far in both directions without intersecting the axes. This phenomenon does not occur when $\alpha < 1$, as is evident in

Fig. 8.1, where all the contours intersect the axes. To verify the phenomenon algebraically, we note that the distribution $\boldsymbol{u} = (1, 1)$ has the same social welfare value as (t, T), where

$$t = \begin{cases} 1/T & \text{if } \alpha = 1 \\ \left(2 - T^{1-\alpha}\right)^{1/(1-\alpha)} & \text{if } \alpha > 1 \text{ and } T^{1-\alpha} < 2 \end{cases}$$

Thus for $\alpha = 1$, we have $t \to 0$ has $T \to \infty$, and for $\alpha > 1$, $t \to 2^{1/(1-\alpha)}$ as $T \to \infty$, even when social welfare is held fixed. If the feasible set is the union of the box $[0, 1] \times [0, 1]$ with the box $[0, t] \times [0, T]$, both $(1, 1)$ and (t, T) are optimal.

8.2 Socially Optimal Distribution

Since the alpha fairness SWF is concave (strictly concave for $\alpha > 0$), classical optimality conditions yield a simple closed-form solution for the optimization problem subject to a budget constraint without utility bounds.

Proposition 8.1 *Maximizing alpha fairness subject to a budget constraint yields an optimal solution in which*

$$u_i^* = \frac{B}{a_i^{1/\alpha} \sum_j a_j^{1-1/\alpha}}, \quad i = 1, \ldots, n$$

In the case of proportional fairness ($\alpha = 1$), each stakeholder i receives utility $u_i^* = B/(na_i)$.

It is evident from Proposition 8.1 that alpha fairness gives some priority to stakeholders with a smaller budget coefficient a_i, but without giving everything to one stakeholder (if $\alpha > 0$) as in a utilitarian context. This is illustrated for three alpha values (0.5, 1, and 2) in Table 2.2, where the budget constraint is $\boldsymbol{au} \le b$ with $\boldsymbol{a} = (5, 6, \ldots, 12)$. Note that as alpha increases, the less efficient stakeholders can receive more resources than the more efficient stakeholders (to compensate for their lower efficiency), although they do not receive a larger utility allotment. Figure 8.4 shows how the distribution changes as α ranges continuously from 0 to 4. As α increases, the solutions transform smoothly from utilitarian to maximin, and the allotments become more egalitarian. The heavy black curve indicates the average utility, which declines somewhat as greater fairness is achieved. Note that after α surpasses 1, the marginal utility cost of greater fairness is practically zero.

Proportional fairness ($\alpha = 1$) is appropriately named because it results in utility allotments that are exactly proportional to efficiency. As a result of this, each stakeholder receives an equal allotment of resources. A solution for alpha fairness can be derived for the case when upper bounds are present, but it is complicated to state and yields little structural insight.

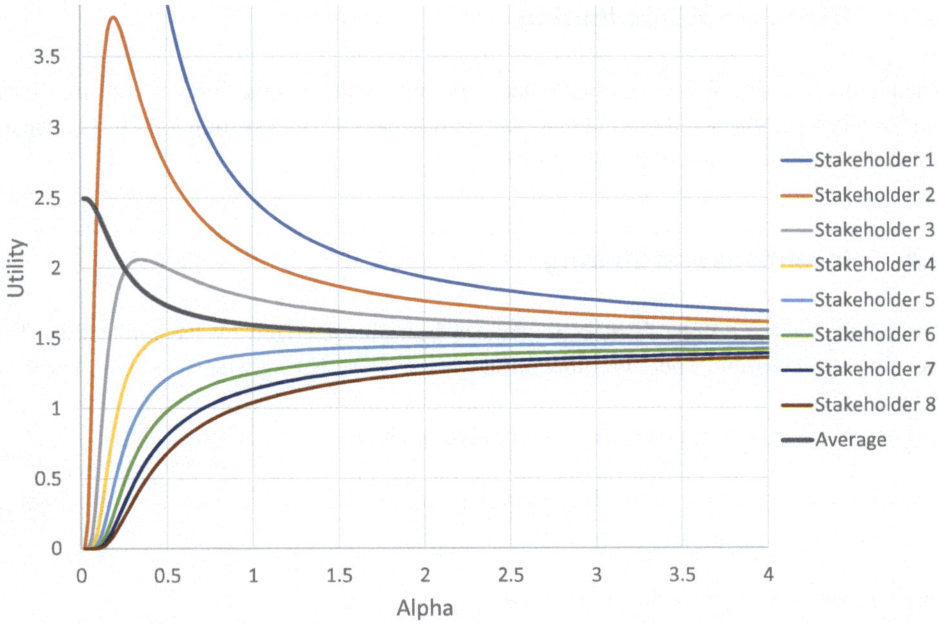

Fig. 8.4 Utility distribution over eight stakeholders using the alpha fairness criterion

Finally, as remarked in Section 2.3, the alpha fairness criterion can model a utility maximizing problem subject to a budget constraint with concave nonlinear utility functions $u_i = x_i^\gamma / a_i$. The latter problem is

$$\max_{\boldsymbol{u},\boldsymbol{x}} \left\{ \sum_i u_i \;\middle|\; u_i = x_i^\gamma / a_i, \text{ all } i; \; \sum_i x_i \le B, \; \boldsymbol{x} \ge \boldsymbol{0} \right\}$$

where $0 < \gamma < 1$. Since $x_i = a_i^{1/\gamma} u_i^{1/\gamma}$, this can be written

$$\max_{\boldsymbol{u}} \left\{ \sum_i u_i \;\middle|\; \sum_i a_i^{1/\gamma} u_i^{1/\gamma} \le B, \; \boldsymbol{u} \ge \boldsymbol{0} \right\}$$

Using the change of variable $\bar{u}_i = u_i^{1/\gamma}$, the problem becomes

$$\max_{\boldsymbol{u}} \left\{ \sum_i \bar{u}_i^\gamma \;\middle|\; \sum_i \bar{a}_i \bar{u}_i \le B, \; \bar{\boldsymbol{u}} \ge \boldsymbol{0} \right\}$$

where $\bar{a}_i = a_i^{1/\gamma}$. But when $0 < \alpha < 1$, this is equivalent to the alpha fairness problem with $\alpha = 1 - \gamma$, because the multiplier $1/(1-\alpha)$ in (8.1) is positive. Also $0 < \alpha < 1$ corresponds to $0 < \gamma < 1$.

8.3 Hierarchical Distribution

The hierarchical problem is regionally decomposable with an alpha fairness criterion since the SWF is separable and therefore monotonically separable (Proposition 3.2). It is collapsible under the condition stated in Proposition 3.1.

8.4 Incentives and Sharing

The parameter α in alpha fairness is perhaps best characterized by its implications for incentives and sharing. There is competition when $\alpha < 1$, sharing when $\alpha > 1$, and neither when $\alpha = 1$.

For a given $\alpha > 0$, the incentive rate for stakeholder i is

$$\frac{1}{\alpha} + \left(1 - \frac{1}{\alpha}\right) \frac{a_i^{1-1/\alpha}}{\sum_j a_j^{1-1/\alpha}} \tag{8.2}$$

and the cross-subsidy rates from i to j are

$$\left(1 - \frac{1}{\alpha}\right) \frac{a_i^{1-1/\alpha}}{\sum_\ell a_\ell^{1-1/\alpha}}, \quad \text{all } j \neq i \tag{8.3}$$

Formula (8.2) implies that for all positive $\alpha \neq 1$, less efficient stakeholders have a stronger incentive to improve efficiency. Formula (8.3) implies that the direction of cross subsidies depends on the size of α. When $\alpha < 1$, efficiency improvements transfer utility from other stakeholders as well as increasing one's own utility creation. This is because the cross-subsidy rate is negative, indicating that the stakeholders are in competition. When $\alpha > 1$, indicating an emphasis on fairness, efficiency improvements benefit other stakeholders as well as oneself. When $\alpha = 1$, corresponding to proportional fairness, the incentive rate is 1, and there are no cross subsidies in either direction.

For the social planner, the technology payoff gradient is a positive scalar multiple of $(a_1^{1-1/\alpha}, \ldots, a_n^{1-1/\alpha})$. This indicates that society should concentrate investment in more efficient stakeholders if $\alpha < 1$, and less efficient stakeholders if $\alpha > 1$, with no distinction between stakeholders if $\alpha = 1$.

References

1. Mo, J., Walrand, J.: Fair end-to-end window-based congestion control. IEEE/ACM Transactions on Networking **8**, 556–567 (2000)
2. Lan, T., Kao, D., Chiang, M., Sabharwal, A.: An axiomatic theory of fairness in network resource allocation. In: Conference on Information Communications (INFOCOM 2010), pp. 1343–1351. IEEE (2010)

3. Bertsimas, D., Farias, V., Trichakis, N.: On the efficiency-fairness trade-off. Management Science **58**, 2234–2250 (2012)
4. Nash, J.: The bargaining problem. Econometrica **18**, 155–162 (1950)
5. Rubinstein, A.: Perfect equilibrium in a bargaining model. Econometrica **50**, 97–109 (1982)
6. Harsanyi, J.C.: Rational Behavior and Bargaining Equilibrium in Games and Social Situations. Cambridge University Press (1977)
7. Binmore, K., Rubinstein, A., Wolinsky, A.: The Nash bargaining solution in economic modeling. RAND Journal of Economics **17**, 176–188 (1986)
8. Lan, T., Kao, D., Chiang, M.: An axiomatic theory of fairness in network resource allocation. In: IEEE INFOCOM (2010)
9. Lan, T., Chiang, M.: An axiomatic theory of fairness in resource allocation. Tech. rep., Princeton University (2011)
10. Kelly, F.P., Maulloo, A.K., Tan, D.K.H.: Rate control for communication networks: Shadow prices, proportional fairness and stability. Journal of the Operational Research Society **49**(3), 237–252 (1998)
11. Mazumdar, R., Mason, L., Douligeris, C.: Fairness in network optimal flow control: Optimality of product forms. IEEE Transactions on Communications **39**(5), 775–782 (1991)
12. Ogryczak, W., Luss, H., Pióro, M., Nace, D., Tomaszewski, A.: Fair optimization and networks: A survey. Journal of Applied Mathematics **2014**, 1–25 (2014)
13. Verloop, I.M., Ayesta, U., Borst, S.: Monotonicity properties for multi-class queueing systems. Discrete Event Dynamic Systems **20**, 473–509 (2010)

Kalai–Smorodinsky Bargaining 9

9.1 Mathematical Formulation

The *Kalai–Smorodinsky* (K–S) bargaining solution can be viewed as maximizing total utility while maintaining fairness for all stakeholders, where fairness takes into account the fact that allocating utility to some stakeholders is more costly than to others. It is defined in terms of the "ideal" utility $u_i^{\max}$ each stakeholder i would have if all available resources were devoted to that stakeholder. The K–S solution then maximizes each stakeholder's fraction of his or her ideal utility, subject to the condition that this fraction is the same for all stakeholders. Viewed in terms of bargaining concessions rather than gains, each stakeholder makes the same relative concession, and the solution minimizes that concession.

Geometrically, the K–S solution can be interpreted as the furthest feasible point from the origin on the line segment connecting the origin and $\boldsymbol{u}^{\max} = (u_1^{\max}, \ldots, u_n^{\max})$. This is illustrated in Fig. 9.1 for two stakeholders, where $\boldsymbol{u}^*$ is the K–S solution, and the feasible set is the portion of the nonnegative quadrant under the curve.

The generic optimization model for n stakeholders is

$$\max_{\lambda,\boldsymbol{u},\boldsymbol{x}} \left\{ \lambda \;\middle|\; \boldsymbol{u} = \lambda \boldsymbol{u}^{\max},\; 0 \le \lambda \le 1,\; (\boldsymbol{u}, \boldsymbol{x}) \in S \right\}$$

where

$$u_i^{\max} = \max_{\boldsymbol{u},\boldsymbol{x}} \left\{ u_i \;\middle|\; (\boldsymbol{u}, \boldsymbol{x}) \in S \right\},\; i = 1, \ldots, n$$

and where $1 - \lambda$ is each stakeholder's relative concession. Unlike alpha fairness, the K–S model offers no parameter to adjust the trade-off between equity and efficiency.

The K–S criterion is naturally seen as applicable to negotiation, as between labor and management [1], who may see a solution as fair when the two parties make the same relative concession. From another perspective, it is appropriate for scenarios where some

Ö. Elçi et al., *The Structure of Fair Solutions*, Synthesis Lectures on Operations Research and Applications, https://doi.org/10.1007/978-3-031-82190-5_9

Fig. 9.1 A Kalai–Smorodinsky bargaining solution for two stakeholders

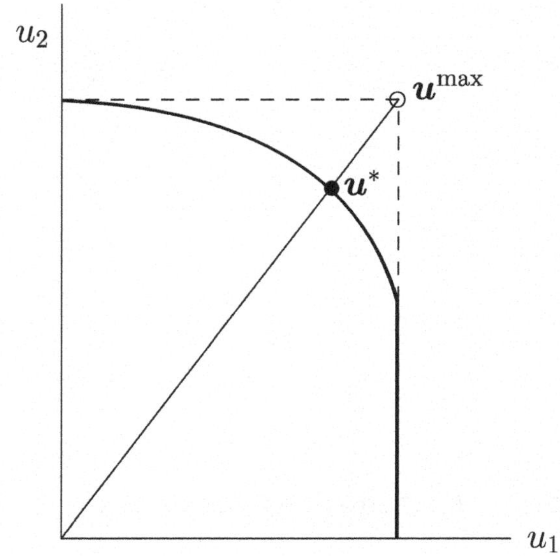

stakeholders are more difficult to accommodate than others. This situation might arise when designing bus routes, since remote customers in suburban and rural areas are more expensive to serve than urban customers.

The K–S criterion was originally proposed as an alternative to the Nash bargaining solution (proportional fairness), in part to avoid a nonmonotonicity property of the latter [2]. Specifically, the Nash solution can allocate *less* utility to some stakeholders when the feasible set is *enlarged*, which could be seen as unacceptable. This is illustrated in Fig. 9.2, where the Nash solution moves from $\boldsymbol{u}^1_{\text{Na}}$ to $\boldsymbol{u}^2_{\text{Na}}$ (indicated by open circles) when the feasible set is enlarged from the region bounded by F_1 to that bounded by F_2. Note that stakeholder 2's share is reduced. This cannot occur when the K–S criterion is used, provided the feasible set is enlarged in such a way that the ideal point $\boldsymbol{u}^{\max}$ is unchanged. This is illustrated in the figure by the two K–S solutions $\boldsymbol{u}^1_{\text{KS}}$ and $\boldsymbol{u}^2_{\text{KS}}$ (black dots).

The K–S criterion has also been defended by Thompson [3], and it is consistent with the contractarian ethical philosophy developed by Gautier [4].

9.2 Socially Optimal Distribution

The K–S solution is easily derived for a budget constraint, with or without upper bounds. Recall that we assume (without loss of generality) that the upper bounds d_i satisfy $d_i \leq B/a_i$ for all i, which means that $\boldsymbol{u}^{\max} = \boldsymbol{d}$. We assume in this section that $\boldsymbol{a}^{\mathsf{T}}\boldsymbol{d} \geq B$, since otherwise, the budget constraint plays no role, and the K–S solution simply sets each utility equal to its upper bound ($\boldsymbol{u} = \boldsymbol{d}$).

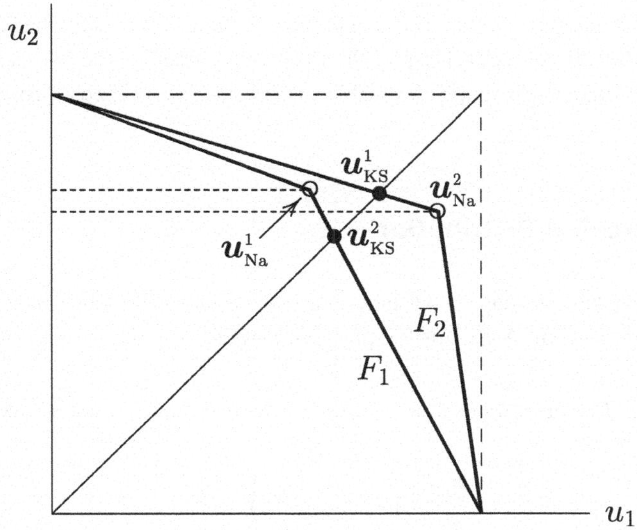

Fig. 9.2 Comparison of Kalai–Smorodinsky solutions (black dots) with Nash solutions (open circles) when the feasible set is enlarged

Proposition 9.1 *If $\boldsymbol{a}^{\mathsf{T}}\boldsymbol{d} \geq B$, then the Kalai–Smorodinsky bargaining solution, subject to a budget constraint and upper bounds $\boldsymbol{d}$, is $\boldsymbol{u}^* = B\boldsymbol{d}/\boldsymbol{a}^{\mathsf{T}}\boldsymbol{d}$. If $\boldsymbol{a}^{\mathsf{T}}\boldsymbol{d} < B$, the solution is $\boldsymbol{u}^* = \boldsymbol{d}$. If there are no upper bounds, the solution is $u_i^* = (1/n)(B/a_i)$ for all i.*

Proof The K–S problem subject to a budget constraint and upper utility bounds is

$$\max_{\boldsymbol{u},\lambda} \left\{\lambda \;\middle|\; \boldsymbol{u} = \lambda \boldsymbol{u}^{\max},\ \boldsymbol{a}^{\mathsf{T}}\boldsymbol{u} \leq B,\ 0 \leq \boldsymbol{u} \leq \boldsymbol{d},\ 0 \leq \lambda \leq 1\right\}$$

Substituting $\lambda \boldsymbol{u}^{\max} = \lambda \boldsymbol{d}$ for $\boldsymbol{u}$, this becomes

$$\max_{\lambda} \left\{\lambda \mid \lambda \boldsymbol{a}^{\mathsf{T}}\boldsymbol{d} \leq B,\ 0 \leq \lambda \boldsymbol{d} \leq \boldsymbol{d},\ 0 \leq \lambda \leq 1\right\}$$

The constraint $0 \leq \lambda\boldsymbol{d} \leq \boldsymbol{d}$ is redundant, and $B/\boldsymbol{a}^{\mathsf{T}}\boldsymbol{d} \leq 1$ since we are given that $\boldsymbol{a}^{\mathsf{T}}\boldsymbol{d} \geq B$. The optimal solution is therefore $\lambda^* = B/\boldsymbol{a}^{\mathsf{T}}\boldsymbol{d} \leq 1$, and $\boldsymbol{u}^* = \lambda^*\boldsymbol{d} = B\boldsymbol{d}/\boldsymbol{a}^{\mathsf{T}}\boldsymbol{d}$. If there are no upper bounds, we can set $d_i = B/a_i$ for each i, so that $\boldsymbol{a}^{\mathsf{T}}\boldsymbol{d} = nB$. Thus we have the optimal solution $u_i^* = (1/n)B/a_i$ for all i.

Thus, in the absence of upper bounds, each stakeholder i receives $1/n$ of its ideal utility B/a_i. This is the same solution as obtained under proportional fairness (alpha fairness with $\alpha = 1$), as exemplified in Table 2.2. Either criterion can allocate far more utility to a stakeholder whose welfare is easily improved (small a_i) than to one who is less fortunate. For example, they may divert treatment resources from cancer patients to persons suffering from the common could in order to provide the two patient groups with the same fraction of their maximum health potential. Despite their leveling tendency, the two criteria can favor stakeholders who happen to enjoy favorable circumstances, perhaps through no merit of their own.

9.3 Hierarchical Distribution

The hierarchical problem may or not be collapsible, and a sufficient condition is given in the following proposition. A collapsible problem is regionally decomposable.

Proposition 9.2 *The hierarchical problem (3.1) with a Kalai–Smorodinsky SWF is collapsible if*

$$\frac{\boldsymbol{a}^k \boldsymbol{d}^k}{\boldsymbol{a}^\mathsf{T} \boldsymbol{d}} \left(B + \sum_{\ell} B_\ell \right) \geq B_k \tag{9.1}$$

for all k. Furthermore, the problem is regionally decomposable if it is collapsible.

Proof The collapsed problem (3.2) is

$$\max_{\boldsymbol{u},\lambda} \left\{ \lambda \;\middle|\; \boldsymbol{u} = \lambda \boldsymbol{d},\; \boldsymbol{a}^\mathsf{T} \boldsymbol{u} \leq B + \sum_{\ell} B_\ell,\; 0 \leq \lambda \leq 1 \right\}$$

By Proposition 9.1, the solution of this problem is

$$\bar{\boldsymbol{u}} = \Big(B + \sum_{\ell} B_i \Big) \boldsymbol{d} / \boldsymbol{a}^\mathsf{T} \boldsymbol{d}, \quad \text{or} \quad \bar{\lambda} = (1/\boldsymbol{a}^\mathsf{T} \boldsymbol{d}) \Big(B + \sum_{\ell} B_\ell \Big) \tag{9.2}$$

where the latter expression is due to $\bar{\boldsymbol{u}} = \bar{\lambda} \boldsymbol{d}$. By Proposition 3.1, this solves the original problem (3.1) if $\boldsymbol{a}^k \bar{\boldsymbol{u}}^k \geq B_k$ for all k. Substituting the value of $\bar{\boldsymbol{u}}$, we obtain (9.1).

To show that the hierarchical problem is regionally decomposable, we note that region k's problem (3.4) is

$$\max_{\lambda_k} \left\{ \lambda_k \mid \boldsymbol{u}^k = \lambda_k \boldsymbol{d}^k,\; \boldsymbol{a}^k \leq B_k + \bar{y}_k,\; 0 \leq \lambda_k \leq 1 \right\}$$

where $\bar{y}_k = \boldsymbol{a}^k \bar{\boldsymbol{u}}^k - B_k$. This simplifies to

$$\max_{\lambda_k} \left\{ \lambda_k \mid \boldsymbol{a}^k \boldsymbol{d}^k \lambda_k \leq \boldsymbol{a}^k \bar{\boldsymbol{u}}^k,\; 0 \leq \lambda_k \leq 1 \right\}$$

Substituting the value of $\bar{\boldsymbol{u}}^k$ in (9.2), this becomes

$$\max_{\lambda_k} \left\{ \lambda_k \;\middle|\; \boldsymbol{a}^k \boldsymbol{d}^k \lambda_k \le \frac{\boldsymbol{a}^k \boldsymbol{d}^k}{\boldsymbol{a}^\mathsf{T} \boldsymbol{d}} \Big(B + \sum_\ell B_\ell \Big),\ 0 \le \lambda_k \le 1 \right\}$$

By Proposition 9.1, the solution of this problem is

$$\hat{\boldsymbol{u}}^k = \hat{\lambda}_k \boldsymbol{d} = \Big(B + \sum_\ell B_\ell \Big) \frac{\boldsymbol{d}}{\boldsymbol{a}^\mathsf{T} \boldsymbol{d}}, \quad \text{or} \quad \hat{\lambda}_k = \frac{1}{\boldsymbol{a}^\mathsf{T} \boldsymbol{d}} \Big(B + \sum_\ell B_\ell \Big)$$

Thus we have $\hat{\lambda}_k = \bar{\lambda}$ for all k, and $\hat{\boldsymbol{u}} = \hat{\lambda}\boldsymbol{d} = \bar{\lambda}\boldsymbol{d} = \bar{u}$ solves the original problem (3.1).

9.4 Incentives and Sharing

The incentive rate, cross subsidies, and technology payoff gradient are the same as for proportional fairness. In particular, changes in a stakeholder's efficiency have no effect on the utility of others. This poses what might be viewed as a classical bargaining situation for independent, self-interested individuals. One's efforts to make better use of resources benefit only oneself, with no necessity to share the surplus with others, and one cannot expect any benefits from the efforts of others.

References

1. Alexander, C.: The Kalai–Smorodinsky bargaining solution in wage negotiations. Journal of the Operational Research Society **43**, 779–786 (1992)
2. Kalai, E., Smorodinsky, M.: Other solutions to Nash's bargaining problem. Econometrica **43**, 513–518 (1975)
3. Thompson, W.: Cooperative models of bargaining. In: R.J. Aumann, S. Hart (eds.) Handbook of Game Theory, vol. 2, pp. 1237–1284. North-Holland (1994)
4. Gautier, D.: Morals by Agreement. Oxford University Press (1983)

Utility Threshold Criterion 10

10.1 Mathematical Formulation

Threshold-based social welfare functions are designed to give priority to stakeholders whose utility level falls below a specified threshold relative to other stakeholders. Williams and Cookson [1] proposed a pair of two-stakeholder SWFs of this sort, one based on a *utility threshold* and one on an *equity threshold*. Both can be generalized to a larger number of stakeholders. They represent very different schemes for combining utilitarian and maximin objectives and may therefore be appropriate for different types of applications. We discuss the utility threshold criterion in this chapter, and the equity threshold criterion in the next.

The threshold criteria studied here are related to but distinct from "sufficientist" SWFs that have been discussed in the economics literature [2, 3], which use a specified minimum utility threshold. They give priority to achieving this minimum before reverting to a purely utilitarian criterion. By contrast, the utility and equity thresholds described here are defined in terms of the relative utility levels of the stakeholders.

A utility threshold criterion is maximin except to the extent that a purely maximin solution sacrifices too much utility, at which point it crosses a utility threshold by treating more privileged stakeholders in a utilitarian manner. The contours of Williams and Cookson's two-person utility threshold SWF are illustrated in Fig. 10.1. The contours are based on a maximin criterion but switch to a utilitarian criterion when $|u_1 - u_2| > \Delta$. If the feasible region is the portion of the nonnegative quadrant under the curve, a maximin solution (small open circle) sacrifices a great deal of utility because stakeholder 2's situation could be much improved. It may therefore be desirable to use a utilitarian solution (solid dot), whose social welfare is slightly greater than that of the maximin solution.

The utility threshold criterion is generalized to n stakeholders in [4], resulting in the following SWF:

Ö. Elçi et al., *The Structure of Fair Solutions*, Synthesis Lectures on Operations Research and Applications, https://doi.org/10.1007/978-3-031-82190-5_10

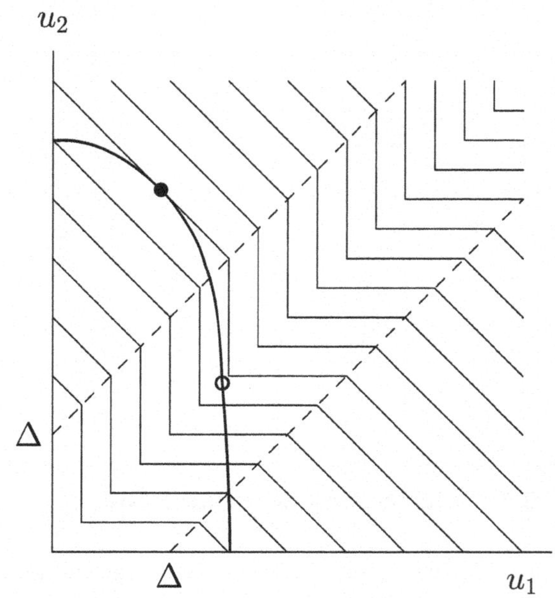

Fig. 10.1 Contours for a two-person utility threshold SWF

$$W_{\Delta}(\boldsymbol{u}) = (n-1)\Delta + \sum_{i=1}^{n} \max\left\{u_i - \Delta,\ u_{\min}\right\} \tag{10.1}$$

where $u_{\min} = \min_i\{u_i\}$. An interpretation of the parameter Δ is proposed in [4] as follows. Utilities within Δ of the lowest utility are regarded as belonging to the *fair region*, and the corresponding stakeholders receive special priority. The remaining stakeholders belong to the *utilitarian region*. The SWF treats utilities in the fair region as though they were equal to the smallest utility, which therefore receives weight in (10.1) equal to the number of utilities in the fair region. Utilities in the utilitarian region receive unit weight. The function becomes purely utilitarian when $\Delta = 0$ and maximin as $\Delta \to \infty$. The parameter Δ is chosen so as to locate utilities in the fair region when the corresponding stakeholders should be seen as disadvantaged enough to deserve higher priority. Larger values of Δ, therefore, place a greater emphasis on fairness as measured by the Rawlsian maximin criterion.

The threshold function $W_{\Delta}(\boldsymbol{u})$ escapes an anomaly that, as noted earlier, characterizes alpha fairness. It cannot assign equality the same social value as arbitrarily extreme inequality. In a two-person context, for example, an egalitarian distribution $\boldsymbol{u} = (1, 1)$ can have the same social value as a distribution in which one party has no utility and the other $\Delta + 2$, but the gap can be no greater than this.

The utility threshold SWF has a mixed integer/linear programming (MILP) model, on the innocuous assumption that there is a number M such that $u_i - u_j \leq M$ for all i, j (which ensures that the SWF is MILP representable). The following is proved in [4]:

Proposition 10.1 *The problem of maximizing $W_\Delta(\boldsymbol{u})$ subject to $(\boldsymbol{u}, \boldsymbol{x}) \in S$ and $u_i - u_j \leq M$ for all i, j is equivalent to the problem:*

$$
\begin{array}{ll}
\max z & \\
z \leq (n-1)\Delta + \sum_i v_i & (a) \\
u_i - \Delta \leq v_i \leq u_i - \delta_i \Delta, \;\; \text{all } i & (b) \\
w \leq v_i \leq w + \delta_i (M - \Delta), \;\; \text{all } i & (c) \\
u_i \geq 0, \;\; \delta_i \in \{0, 1\}, \;\; \text{all } i & \\
(\boldsymbol{u}, \boldsymbol{x}) \in S &
\end{array}
\tag{10.2}
$$

Proof The problem of maximizing $W_\Delta(\boldsymbol{u})$ subject to $(\boldsymbol{u}, \boldsymbol{x}) \in S$ can be written as

$$
\max_{\boldsymbol{u}, \boldsymbol{x}, z} \left\{ z \;\middle|\; z \leq (n-1)\Delta + \sum_i \max\left\{u_{\min}, u_i - \Delta\right\}, \;\; (\boldsymbol{u}, \boldsymbol{x}) \in S \right\} \tag{10.3}
$$

It suffices to show that for any feasible solution of (10.3), there is a feasible solution of (10.2) with the same z-value, and vice versa. To show the former, consider any feasible solution $(\boldsymbol{u}, \boldsymbol{x}, z)$ of (10.3). Then $(\boldsymbol{u}, \boldsymbol{x}, z, \boldsymbol{v}, w, \boldsymbol{\delta})$ is a feasible solution of (10.2), where

$$
w = u_{\min}, \quad (\delta_i, v_i) = \begin{cases} (0, u_{\min}), & \text{if } u_i - u_{\min} < \Delta \\ (1, u_i - \Delta), & \text{otherwise} \end{cases} \tag{10.4}
$$

If $u_i - u_{\min} < \Delta$, then $u_i - \Delta < u_{\min}$, and it is easily checked that constraints (a)–(c) are satisfied. If $u_i - u_{\min} \geq \Delta$, then $u_i - \Delta \geq u_{\min}$, which implies (a), (b), and the first inequality of (c). The second inequality of (c) is satisfied because $u_i - u_j \leq M$ for all i, j.

We now suppose that $(\boldsymbol{u}, \boldsymbol{x}, z, \boldsymbol{v}, w, \boldsymbol{\delta})$ is feasible in (10.2) and show that $(\boldsymbol{u}, \boldsymbol{x}, z)$ satisfies the constraints of (10.3) as well as $u_i - u_j \leq M$ for all i, j. To show the latter, note that (c) implies $v_j - (M - \Delta)\delta_j \leq w \leq v_i$ for any i, j, and therefore $v_j - v_i \leq M - \Delta$. But this implies $u_j - u_i \leq M$ because $v_j \geq u_j - \Delta$ and $v_i \leq u_i$ due to (b). To show that $(\boldsymbol{u}, \boldsymbol{x}, z)$ is feasible in (10.2), we write (a) as

$$
z \leq (n-1)\Delta + \sum_{i \mid \delta_i = 0} v_i + \sum_{i \mid \delta_i = 1} v_i \tag{10.5}
$$

Consider any term v_i in the first summation. We have from (b) that $v_j \leq u_j - \delta_i \Delta$ for all j, which implies $v_i \leq u_{\min}$. Thus $v_i \leq \max\{u_{\min}, u_i - \Delta\}$. For any term v_i in the second summation of (10.5), we have from (b) that $v_i = u_i - \Delta$, and again $v_i \leq \max\{u_{\min}, u_i - \Delta\}$. It follows that the bound on z in (10.5) implies the bound on z in (10.3), and $(\boldsymbol{u}, \boldsymbol{x}, z)$ is therefore feasible in (10.3).

The utility threshold criterion is extended in [4] to accommodate differently sized groups of stakeholders as well as individual stakeholders. The utility u_i associated with a group is taken to be the average utility of the group for purposes of assessing fairness, and the total utility of the group for purposes of computing overall utility. Thus the trade-off of fairness and efficiency remains, but with a focus on fairness across groups rather than across individuals. If s_i is the size of group i, the SWF is

$$W_\Delta(\boldsymbol{u}) = \Big(\sum_i s_i - 1\Big)\Delta + \sum_i s_i \max\big\{u_{\min}, u_i - \Delta\big\}$$

An MILP model of the group-oriented problem is as follows. The proof of validity, given in [4], is very similar to the proof of Proposition 10.1.

$$\begin{aligned}
&\max z \\
&z \le \Big(\sum_i s_i - 1\Big)\Delta + \sum_i s_i v_i \\
&u_i - \Delta \le v_i \le u_i - \delta_i \Delta, \quad \text{all } i \\
&w \le v_i \le w + \delta_i (M - \Delta), \quad \text{all } i \\
&u_i \ge 0, \ \delta_i \in \{0, 1\}, \quad \text{all } i \\
&(\boldsymbol{u}, \boldsymbol{x}) \in S
\end{aligned} \tag{10.6}$$

It is shown in [4] that both of the models (10.2) and (10.6) are sharp [5], apart from the constraints $(\boldsymbol{u}, \boldsymbol{x}) \in S$. A somewhat simpler proof appears in [6].

10.2 Socially Optimal Distribution

When the utility-based threshold SWF is maximized subject to a budget constraint, the solution is either purely utilitarian or purely maximin, depending on the value of Δ. Let $\boldsymbol{e}_i$ be the ith unit vector, and let $\boldsymbol{e}$ be a vector of ones. Then we have

Proposition 10.2 *Assume the stakeholders are indexed so that $a_1 \le \cdots \le a_n$. Then maximizing the utility threshold SWF subject to a budget constraint yields a purely utilitarian optimal solution $\boldsymbol{u}^* = (B/a_1)\boldsymbol{e}_1$ when*

$$\Delta \le B\left(\frac{1}{a_1} - \frac{n}{a(n)}\right) \tag{10.7}$$

and otherwise a purely maximin solution $\boldsymbol{u}^* = (B/a(n))\boldsymbol{e}$.

To prove Proposition 10.2, it is useful to reformulate (2.6) as follows:

$$\max_{v_0, \boldsymbol{v}} \left\{\overline{W}(v_0, \boldsymbol{v}) \;\middle|\; a(n)v_0 + \boldsymbol{a}^\top \boldsymbol{v} \le B, \ \boldsymbol{v} \ge \boldsymbol{0}\right\} \tag{10.8}$$

where $\boldsymbol{v} = (v_1, \ldots, v_n)$ and

$$\overline{W}(v_0, \boldsymbol{v}) = (n-1)\Delta + n v_0 + \sum_{i=1}^{n} (v_i - \Delta)^+$$

Lemma 10.1 *Formulation* (10.8) *has the same optimal value as (2.6).*

Proof It suffices to show that for any feasible solution of (2.6), there is a feasible solution of (10.8) with a value at least as large as that of (2.6), and vice versa. First consider any feasible solution $\boldsymbol{u}$ of (2.6). If we let $v_0 = u_{\min}$ and $v_i = u_i - u_{\min}$ for all i, the solution $(v_0, \boldsymbol{v})$ is feasible in (10.8), given the constraints of (2.6). Also the objective function $\overline{W}$ is identical to W, and so $\overline{W}(v_0, \boldsymbol{v}) = W(\boldsymbol{u})$.

Now suppose that $(v_0, \boldsymbol{v})$ is feasible in (10.8). Set $u_i = v_0 + v_i$ for all i, which implies $u_{\min} = v_0 + v_{\min}$, where $v_{\min} = \min_i\{v_i\}$. The constraint $\boldsymbol{a}^{\mathsf{T}}\boldsymbol{u} \leq B$ of (2.6) becomes $a(n)v_0 + \boldsymbol{a}^{\mathsf{T}}\boldsymbol{v} \leq B$ in (10.8), so that $\boldsymbol{u}$ is feasible in (2.6). The objective function of (2.6) becomes

$$(n-1)\Delta + nv_{\min} + nv_0 + \sum_i \left(v_i - v_{\min} - \Delta\right)^+$$

which can be written as

$$(n-1)\Delta + nv_0 + \sum_i \max\left\{v_i - \Delta,\ v_{\min}\right\}$$

This is no smaller than the objective function of (10.8) because $v_{\min} \geq 0$.

Proof of Proposition 10.2. We will show that $(v_0^*, \boldsymbol{v}^*) = (0, (B/a_1)\boldsymbol{e}_1)$ is an optimal solution of (10.8) if (10.7) holds, and $(v_0^*, \boldsymbol{v}^*) = (B/a(n), \boldsymbol{0})$ is an optimal solution otherwise. This proves the theorem because (2.6) and (10.8) have the same optimal value by Lemma 10.1, and because

$$W\Big(\frac{B}{a_1}\boldsymbol{e}_1\Big) = (n-2)\Delta + \frac{B}{a_1} = \overline{W}\Big(0, \frac{B}{a_1}\boldsymbol{e}_1\Big)$$
$$W\Big(\frac{B}{a(n)}\boldsymbol{e}\Big) = (n-1)\Delta + \frac{nB}{a(n)} = \overline{W}\Big(\frac{B}{a(n)}, \boldsymbol{0}\Big)$$

We first observe that the objective function $\overline{W}(v_0, \boldsymbol{v})$ is convex because $(v_i - \Delta)^+$ is a convex function of v_i, and a sum of convex functions is convex. It follows that some extreme point of the feasible set of (10.8) is optimal. Yet every extreme point is the solution of some linearly independent subset T of $n+1$ of the following equations:

$$\begin{aligned} &a(n)v_0 + \boldsymbol{a}\boldsymbol{v} = B && (a)\\ &v_0 = 0 && (b)\\ &v_i = 0,\ i = 1, \ldots, n && (c) \end{aligned}$$

We can suppose T contains (a), since otherwise the corresponding extreme point $(v_0, \boldsymbol{v}) = (0, \boldsymbol{0})$ is clearly dominated by $(0, (B/a_1)\boldsymbol{e}_1)$ and $(B/a(n), \boldsymbol{0})$. Then T either contains (b) and all but one equation $v_t = 0$ in (c), or else all equations in (c). The former yields extreme point $(v_0, \boldsymbol{v}) = (0, (B/a_t)\boldsymbol{e}_t)$ and the latter $(v_0, \boldsymbol{v}) = (B/a(n), \boldsymbol{0})$. Now if (10.7) holds, both $\overline{W}(0, (B/a_j)\boldsymbol{e}_t) = (n-2)\Delta + B/a_t$ and $\overline{W}(B/a(n), \boldsymbol{0}) = (n-1)\Delta + nB/a(n)$ are less than or equal to $\overline{W}(0, (B/a_t)\boldsymbol{e}_t)$ because $a_1 \le a_t$. If (10.7) does not hold, both of these expressions are less than or equal to $\overline{W}(B/a(n), \boldsymbol{0})$. The proposition follows. □

In the example of Table 2.2, the critical value of Δ is 8.235. When there are lower and upper bounds on stakeholder utilities, as often occurs in practice, the solutions are less extreme and perhaps more useful. They have an interesting structure as well.

Proposition 10.3 *Maximizing a utility threshold SWF subject to a budget constraint and upper and lower utility bounds yields an optimal solution in which at most one utility u_i is strictly between $u_{\min}$ and d_i. Furthermore, if there is such a utility, then some other utility u_j that is equal to $u_{\min}$ is at its lower or upper bound.*

Proof We first show that maximizing a utility threshold SWF subject to a budget constraint and upper and lower utility bounds yields an optimal solution $\boldsymbol{u}^*$ in which at most one utility u_i^* is strictly between $u_{\min}^*$ and d_i.

Let S be the feasible set of (2.4), and define

$$S_i = S \cap \{\boldsymbol{u} \mid u_i \le u_j, \text{ all } j\}$$

Since S is the union of all S_i, the maximum of $W_\Delta(\boldsymbol{u})$ over some S_i is optimal in (2.4). Suppose without loss of generality that the maximum of $W_\Delta(\boldsymbol{u})$ over S_1 is optimal in (2.4). For $\boldsymbol{u} \in S_1$, the function $W_\Delta(\boldsymbol{u})$ can be written as

$$W'_\Delta(\boldsymbol{u}) = (n-1)\Delta + \sum_{i=1}^{n} \max\left\{u_i - \Delta,\ u_1\right\}$$

Since $W'_\Delta(\boldsymbol{u})$ is convex, some extreme point $\boldsymbol{u}^*$ of S_1 maximizes $W'_\Delta(\boldsymbol{u})$ and therefore $W_\Delta(\boldsymbol{u})$ over S_1. Since $\boldsymbol{u}^*$ is an extreme point of S_1, it is the solution of some linearly independent[1] subset E of n of the equations:

$$\begin{aligned}
&u_1 - u_j = 0,\ j = 2, \ldots, n && (a)\\
&\boldsymbol{a}^{\mathsf{T}}\boldsymbol{u} = B && (b)\\
&\boldsymbol{u} = \boldsymbol{c} && (c)\\
&\boldsymbol{u} = \boldsymbol{d} && (d)
\end{aligned}$$

Let T be the subset of equations in E that appear in (c) or (d). This means that $n - |T|$ variables are not fixed to one of their bounds by (c) and (d). Suppose that 2 or more of these variables are not set equal to u_1 by (a). Then the only nonzero in each of the corresponding columns must appear in (b). These columns must therefore be linearly dependent, which

[1] We consider equations to be linearly independent when their coefficient rows are linearly independent.

is impossible because E is nonsingular. We conclude that at most one variable u_i^* is fixed neither to a bound by (c) and (d) nor to $u_1^* = u_{\min}^*$ by (a). Since u_i^* cannot be strictly between its lower bound c_i and $u_{\min}^*$, the proposition follows.

We next show that if there is utility strictly between $u_{\min}^*$ and its upper bound, then some other utility u_j^* that is equal to $u_{\min}^*$ is at its lower or upper bound. Let Q be the set of inequalities (a) in E. Then Q contains $|Q| + 1$ variables, and T contains $|T|$ variables. Since m variables are strictly between $u_{\min}^*$ and their upper bounds, these variables are in neither Q nor T, and E must therefore contain all m rows of (b). If Q and T have no variables in common, the number of variables is at least $|Q| + |T| + m + 1$. But $|Q| + |T| + m + 1 > n$ because $|Q| + |T| + m = n$ due to the linear independence of the equations in E. Thus one variable u_j^* belongs to both Q and T, which implies that $u_j^* = u_1^* = u_{\min}^*$ and u_j^* is at its lower or upper bound.

This result says that nearly all utilities will be at their upper bound or equal to the lowest utility. It is illustrated for two persons in Fig. 10.2, where u_1^* is strictly between the upper bound d_1 and $u_{\min}^*$, but $u_2^* = u_{\min}^*$. This kind of structure can simplify implementation and provide managerial insight. In the healthcare example of Section 2.2, it tells us that nearly all patients (all but those suffering from one particular disease) will either receive their maximum possible utility or else end up as one of the worst-off patients, who are given the highest priority.

10.3 Hierarchical Distribution

We can specialize Proposition 10.3 to a hierarchical network with r regions, each of which has subregions.

Proposition 10.4 *Any optimal solution $\bar{\boldsymbol{u}}$ of (3.1) with a utility-based threshold function $W_\Delta(\boldsymbol{u})$ satisfies the following:*

(i) At most r utilities lie strictly between $\bar{u}_{\min}$ and their upper bounds.
(ii) At most one utility in any region lies strictly between $\bar{u}_{\min}$ and its upper bound.

Proof Let S be the feasible set of (3.1), and define

$$S_i = S \cap \{(\boldsymbol{y}, \boldsymbol{u}) \mid u_i \le u_j, \text{ all } j\}$$

Since S is the union of all S_i, the maximum of $W_\Delta(\boldsymbol{u})$ over some S_i is optimal in (3.1). Suppose without loss of generality that the maximum of $W_\Delta(\boldsymbol{u})$ over S_1 is optimal in (3.1). For $(\boldsymbol{y}, \boldsymbol{u}) \in S_1$, the function $W_\Delta(\boldsymbol{u})$ can be written as

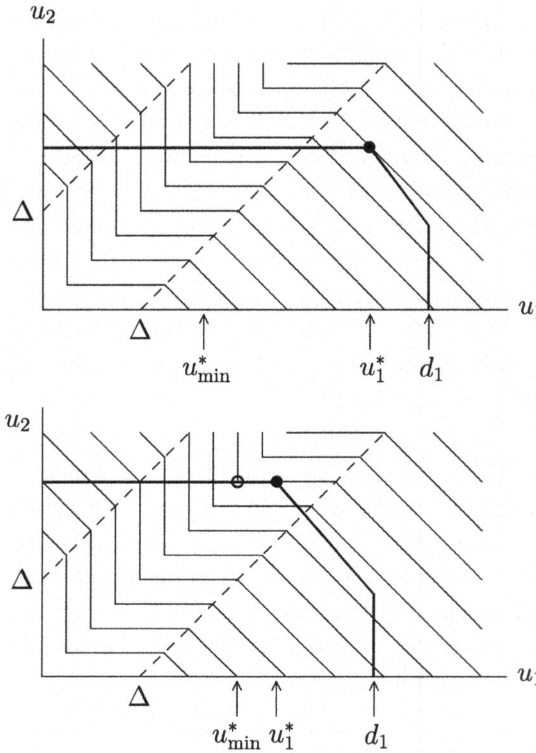

Fig. 10.2 Two instances of the utility threshold problem in which u_1^* lies strictly between d_1 and u^*_{min} in an optimal solution (u_1^*, u_2^*) (black dot). Note that in the second instance, there is another optimal solution in which $u_1^* = u^*_{\text{min}}$ (open circle)

$$W_1(\boldsymbol{u}) = (n-1)\Delta + nu_1 + \sum_{i=1}^{n} \left(u_i - u_1 - \Delta\right)^+$$

Since $W_1(\boldsymbol{u})$ is convex, some extreme point $(\bar{\boldsymbol{y}}, \bar{\boldsymbol{u}})$ of S_1 maximizes $W_1(\boldsymbol{u})$ and therefore $W_\Delta(\boldsymbol{u})$ over S_1. Since $(\bar{\boldsymbol{y}}, \bar{\boldsymbol{u}})$ is an extreme point of S_1, it is the solution of some linearly independent subset E of $n + r$ of the equations:

$$\begin{array}{ll}
u_1 - u_j = 0, \;\; j = 2, \ldots, n & (a) \\
\boldsymbol{e}^{\mathsf{T}} \boldsymbol{y} = B & (b) \\
\boldsymbol{a}^k \boldsymbol{u} - y_k = B_k, \;\; k = 1, \ldots, r & (c) \\
u_i = c_i, \;\; i = 1, \ldots, n & (d) \\
u_i = d_i, \;\; i = 1, \ldots, n & (e) \\
y_k = 0, \;\; k = 1, \ldots, r & (f)
\end{array}$$

We first demonstrate (i) Let T be the subset of equations in E that appear in (d) or (e). This means that $n - |T|$ variables $\bar{u}_j$ are not fixed to one of their bounds by (d) and (e). Suppose that $r + 1$ of these variables are not set equal to $\bar{u}_1$ by (a). Then all of the nonzeros in the corresponding columns must occur in the r rows of (c). The $r + 1$ columns must therefore be linearly dependent, which is impossible because is E is nonsingular. We conclude that at most r variables $\bar{u}_j$ are fixed neither to a bound by (d) and (e) nor to $\bar{u}_1 = \bar{u}_{\min}$ by (a). Since these $\bar{u}_j$s cannot be strictly between their lower bound c_j and $\bar{u}_{\min}$, (i) follows.

We now demonstrate (ii) Let T be as before. This means that $n - |T|$ variables $\bar{u}_j$ are not fixed to one of their bounds by (d) and (e). Suppose for any given k that two of these variables that are in $\bar{\boldsymbol{u}}^k$ are not set equal to $\bar{u}_1$ by (a). Then all of the nonzeros in the corresponding columns must occur in row k of (c). The two columns must therefore be linearly dependent, which is impossible because E is nonsingular. We conclude that at most one variable $\bar{u}_j$ in $\bar{\boldsymbol{u}}^k$ is strictly between its lower bound c_j and $\bar{u}_{\min}$, and (ii) follows.

The threshold function $W_\Delta(\boldsymbol{u})$ is not monotonically separable, and so there is no assurance that a given instance of the problem is regionally decomposable. As a simple example, suppose there are two regions, one with subregional utilities u_1, u_2 and the other with a single utility u_3, and let $\Delta = 1$. Then if $\boldsymbol{a} = (1, 1, 4)$ and $(B, B_1, B_2) = (1, 1, 0)$, Proposition 10.2 tells us that the solution of the collapsed problem (3.2) is $\boldsymbol{u} = (2, 0, 0)$. This instance of the problem is, in fact, collapsible by Proposition 3.1 because $(B_1, B_2) \leq (2, 0)$, and so $\boldsymbol{u} = (2, 0, 0)$ solves the original problem (3.1). However, the regionally optimal solution for (u_1, u_2) is $(1, 1)$, which is suboptimal in the national problem. This instance of the problem is therefore not regionally decomposable.

10.4 Incentives and Sharing

The benefits of improving one's efficiency are distributed as in a purely utilitarian or purely maximin solution, depending on which one obtains for the chosen value of Δ.

References

1. Williams, A., Cookson, R.: Equity in Health. Culyer, A.J. and Newhouse, J.P. editors, Handbook of Health Economics (2000)
2. Adler, M.D.: Measuring Social Welfare. Oxford University Press (2019)
3. Crisp, R.: Equality, priority, and compassion. Ethics **113**, 745–763 (2003)
4. Hooker, J.N., Williams, H.P.: Combining equity and utilitarianism in a mathematical programming model. Management Science **58**, 1682–1693 (2012)
5. Vielma, J.P.: Mixed integer linear programming techniques. SIAM Review **57**, 3–57 (2015)
6. Chen, V., Hooker, J.N.: Combining leximax fairness and efficiency in an optimization model. European Journal of Operational Research **299**, 235–248 (2022)

11 Equity Threshold Criterion

11.1 Mathematical Formulation

An equity threshold criterion is utilitarian until it begins to create excessive inequality, at which point it crosses an equity threshold and gives additional priority to less privileged stakeholders. Contours of the two-person equity threshold SWF of [1] are illustrated in Fig. 11.1. They are based on a utilitarian function but switch to a maximin criterion when $|u_1 - u_2| > \Delta$. The SWF can be generalized to n persons as follows:

$$W_\Delta(\boldsymbol{u}) = n\Delta + \sum_i \min\left\{u_i - \Delta, u_{\min}\right\} \tag{11.1}$$

The interpretation of Δ is different than for the utility threshold function. Increasing a utility u_i that already exceeds the lowest by more than Δ is viewed as adding nothing to social welfare, unless the lowest is increased an equal amount. Thus larger values of Δ correspond to more nearly utilitarian solutions, rather than smaller values as in the utility threshold problem. In particular, $\Delta = 0$ yields a pure maximin solution, and sufficiently large Δ yields a pure utilitarian solution.

The analysis of the equity threshold SWF is also quite different because, unlike the utility threshold SWF, it is concave. In fact, it has a linear model. If the problem of maximizing (11.1) subject to $(\boldsymbol{u}, \boldsymbol{x}) \in S$ is written as

$$\max_{\boldsymbol{u},\boldsymbol{x}} \left\{ n\Delta + \sum_i \min\left\{u_i - \Delta, u_{\min}\right\} \;\middle|\; (\boldsymbol{u}, \boldsymbol{x}) \in S \right\} \tag{11.2}$$

then we have

Ö. Elçi et al., *The Structure of Fair Solutions*, Synthesis Lectures on Operations Research and Applications, https://doi.org/10.1007/978-3-031-82190-5_11

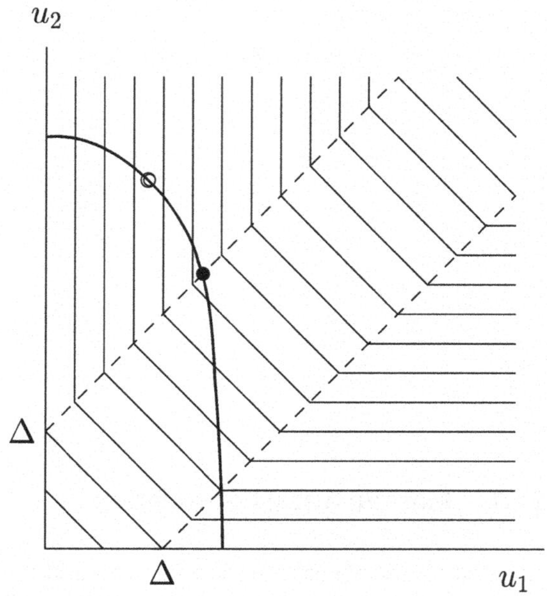

Fig. 11.1 Contours for a two-person equity threshold SWF. The black dot indicates an optimal solution for the equity threshold criterion, and the open circle a utilitarian solution

Proposition 11.1 *The optimization problem* (11.2) *can be formulated*

$$\max_{\boldsymbol{u},\boldsymbol{x},\boldsymbol{v},w} \left\{ n\Delta + \sum_i v_i \;\middle|\; \begin{array}{l} w \leq u_i,\ v_i \leq u_i - \Delta,\ v_i \leq w,\ i = 1, \ldots n \\ w \geq 0,\ (\boldsymbol{u}, \boldsymbol{x}) \in S \end{array} \right\} \tag{11.3}$$

where $\boldsymbol{v}$ *and* w *are auxiliary variables.*

Proof It suffices to show that for any feasible solution of (11.3), some feasible solution of (11.2) has an objective function value at least as large, and vice versa. For the first claim, if we let $(\boldsymbol{u}, \boldsymbol{x}, \boldsymbol{v}, w)$ be a feasible solution of (11.3), then $(\boldsymbol{u}, \boldsymbol{x})$ is obviously feasible in (11.2), and it suffices to show that

$$n\Delta + \sum_i \min\left\{u_i - \Delta,\ u_{\min}\right\} \geq n\Delta + \sum_i v_i$$

But this follows because $v_i \leq u_i - \Delta$ and $v_i \leq w \leq u_{\min}$ for each i, the latter due to the fact that $w \leq u_i$ for each i. To show the converse, suppose $\boldsymbol{u}$ is a feasible solution of (11.2). It suffices to exhibit a feasible solution $(\boldsymbol{u}, \boldsymbol{v}, w)$ of (11.3) for which

$$n\Delta + \sum_i v_i \geq n\Delta + \sum_i \min\left\{u_i - \Delta,\ u_{\min}\right\} \tag{11.4}$$

The solution given by $w = u_{\min}$ and $v_i = \min\{u_i - \Delta, u_{\min}\}$ for each i is feasible on inspection. We also see that (11.4) follows immediately from the definition of v_i.

Like the utility threshold SWF, the equity threshold SWF can be extended to accommodate groups of stakeholders as well as individual stakeholders. If s_i is the size of group i, the SWF becomes

$$W_\Delta(\boldsymbol{u}) = n\Delta + \sum_i \min\left\{u_i - \Delta, u_{\min}\right\}$$

A linear programming formulation of the optimization is (11.3) with the objective function replaced by

$$\Big(\sum_i s_i\Big)\Delta + \sum_i s_i v_i$$

The validity proof of this formulation is very similar to the proof of Proposition 11.1.

11.2 Socially Optimal Distribution

The welfare maximizing solution has a more complex (and perhaps more interesting) structure for an equity threshold function than for a utility threshold function. When maximizing welfare subject to a budget constraint, values of Δ larger than a critical value Δ^* result in an allocation of utility Δ to a few of the more efficient stakeholders, while all other stakeholders (with one exception in general) receive nothing. As Δ increases, fewer stakeholders benefit, with only one stakeholder receiving utility in the pure utilitarian case. Values of Δ smaller than Δ^* allocate utility to everyone, but at two levels. Less efficient stakeholders receive utility u_0 and more efficient stakeholders receive utility $u_0 + \Delta$, where u_0 depends on the problem data. In the example of Table 2.2, $\Delta^* = 3.846$. The precise result is stated in the following theorem, whose proof uses the fact that the optimization problem has a linear model:

Proposition 11.2 *Suppose $a_1 \leq \cdots \leq a_n$, and let k^* be the largest k for which $a(n)/a_k > n$. Then for $\Delta < \Delta^*$ where*

$$\Delta^* = \frac{B}{a(k^*)} \tag{11.5}$$

an optimal solution subject to a budget constraint for the equity threshold criterion is given by

$$u_i^* = \begin{cases} u_0 + \Delta, & i = 1, \ldots, k^* \\ u_0, & i = k^* + 1, \ldots, n \end{cases} \tag{11.6}$$

where

$$u_0 = \frac{1}{a(n)}\Big(B - \Delta a(k^*)\Big) \tag{11.7}$$

For $\Delta \geq \Delta^*$, an optimal solution is given by

$$u_i^* = \begin{cases} \Delta, & i = 1, \ldots, k-1 \\ \dfrac{1}{a_k}\Big(B - \Delta a(k-1)\Big), & i = k \\ 0, & i = k+1, \ldots, n \end{cases} \tag{11.8}$$

when $k \le k^*$ *and* k *satisfies*

$$\frac{B}{a(k)} < \Delta \le \frac{B}{a(k-1)}$$

where $a(0) = 0$ *by convention.*

Proof Due to Proposition 11.1, the optimization problem subject to a linear budget constraint can be written as

$$\min_{\alpha,\beta,\gamma,\delta} \left\{ n\Delta + B\delta - \Delta\sum_{i=1}^{n}\beta_i \;\middle|\; \begin{array}{l} a_i\delta \ge \alpha_i + \beta_i,\; \beta_i + \gamma_i = 1,\; i \in N \\ \sum_i \alpha_i \ge \sum_i \gamma_i,\; \boldsymbol{\alpha}, \boldsymbol{\beta}, \boldsymbol{\gamma} \ge \mathbf{0},\; \delta \ge 0 \end{array} \right\} \tag{11.9}$$

where $\boldsymbol{v}$ and w are auxiliary variables. It therefore suffices to show that (11.6) and (11.8) hold under their respective conditions for some optimal solution $\boldsymbol{u}^*$ of (11.9). Since (11.9) is an LP, we can show this by exhibiting a feasible solution $(\boldsymbol{u}^*, \boldsymbol{v}^*, w^*)$ of (11.9) that is consistent with (11.6) or (11.8), along with a feasible dual solution that yields the same objective function value as $(\boldsymbol{u}^*, \boldsymbol{v}^*, w^*)$. The dual of (11.9) is

$$\min_{\alpha,\beta,\gamma,\delta} \left\{ n\Delta + B\delta - \Delta\sum_{i=1}^{n}\beta_i \;\middle|\; \begin{array}{l} a_i\delta \ge \alpha_i + \beta_i,\; \beta_i + \gamma_i = 1,\; i \in N \\ \sum_i \alpha_i \ge \sum_i \gamma_i,\; \boldsymbol{\alpha}, \boldsymbol{\beta}, \boldsymbol{\gamma} \ge \mathbf{0},\; \delta \ge 0 \end{array} \right\} \tag{11.10}$$

where dual variables $\boldsymbol{\alpha}, \boldsymbol{\beta}, \boldsymbol{\gamma}, \delta$ correspond, respectively, to the first four constraints of (11.9). We first suppose $\Delta < \Delta^*$. It suffices to show that the optimal solution $(\boldsymbol{u}^*, \boldsymbol{v}^*, w^*)$ is optimal when $\boldsymbol{u}^*$ is as given by (11.6), $\boldsymbol{v}^*$ is given by

$$v_i^* = \begin{cases} u_0, & \text{for } i = 1, \ldots, k^* \\ u_0 - \Delta, & \text{for } i = k^*+1, \ldots, n \end{cases}$$

and $w^* = u_0$. We note that $\Delta < \Delta^*$ implies $u_0 \ge 0$. Given this, it is easily checked that $(\boldsymbol{u}^*, \boldsymbol{v}^*, w^*)$ is feasible in (11.9). To show it is optimal, we exhibit the dual solution $(\boldsymbol{\alpha}, \boldsymbol{\beta}, \boldsymbol{\gamma}, \delta)$ given by

$$(\alpha_i, \beta_i) = \begin{cases} (0, \dfrac{na_j}{a(n)}), & \text{for } i = 1, \ldots, k^* \\ \Big(\dfrac{na_j}{a(n)} - 1,\ 1\Big), & \text{for } i = k^*+1, \ldots, n \end{cases}$$

as well as $\gamma_i = 1 - \beta_i$ for $i = 1, \ldots, n$ and $\delta = n/a(n)$. We note that $\alpha_i \ge 0$ because $na_i/a(n) \ge 1$ for $i = k^*+1, \ldots, n$ due to the definition of k^*. Given this, direct substitution confirms that $(\boldsymbol{\alpha}, \boldsymbol{\beta}, \boldsymbol{\gamma}, \delta)$ is dual feasible. Furthermore, the objective value of both primal and dual is

$$k\Delta + \frac{n}{a(n)}\Big(B - a(k^*)\Delta\Big)$$

This shows that $(\boldsymbol{u}^*, \boldsymbol{v}^*, w^*)$ is optimal.

We now suppose $\Delta \geq \Delta^*$. It suffices to show that the optimal solution $(\boldsymbol{u}^*, \boldsymbol{v}^*, w^*)$ is optimal when $\boldsymbol{u}^*$ is given by (11.8) for values of $k \leq k^*$ satisfying (11.7), when $v_i^* = u_i^* - \Delta$ for all i, and when $w^* = 0$. We note that $u_k^* \geq 0$ due to the second inequality in (11.7). Given this, it is easily checked that $(\boldsymbol{u}^*, \boldsymbol{v}^*, w^*)$ is feasible in (11.3). To show it is optimal, we exhibit the dual solution $(\boldsymbol{\alpha}, \boldsymbol{\beta}, \boldsymbol{\gamma}, \delta)$ given by

$$(\alpha_i, \beta_i) = \begin{cases} \left(0, \ \dfrac{a_i}{a_k}\right), & \text{for } i = 1, \ldots, k \\ \left(\dfrac{a_i}{a_k} - 1, \ 1\right), & \text{for } i = k+1, \ldots, n \end{cases}$$

as well as $\gamma_i = 1 - \beta_i$ for $i = 1, \ldots, n$ and $\delta = 1/a_k$. We note that $\alpha_i \geq 0$ because $a_i \geq a_k$ for $i = k+1, \ldots, n$. We also note that the dual constraint $\sum_i \alpha_i \geq \sum_i \gamma_i$ reduces to $a_k \leq a(n)/n$, which holds because $k \leq k^*$ and the definition of k^* implies $a_{k^*} \leq a(n)/n$. Given these facts, direct substitution confirms that $(\boldsymbol{\alpha}, \boldsymbol{\beta}, \boldsymbol{\gamma}, \delta)$ is dual feasible. Furthermore, the objective value of both primal and dual is

$$(k-1)\Delta + \frac{1}{a_k}\Big(B - a(k{-}1)\Delta\Big)$$

This shows that $(\boldsymbol{u}^*, \boldsymbol{v}^*, w^*)$ is optimal.

As an example, suppose we wish to maximize the equity threshold SWF when there are four stakeholders' budget constraint $3u_1 + 4u_2 + 7u_3 + 12u_4 \leq B$ and $\Delta = 2$. If $B = 12$, we have $k^* = 2$ and $\Delta^* = 1\frac{5}{7}$. Since $\Delta \geq \Delta^*$ and $k = 2$, the solution is $\boldsymbol{u} = (2, 1\frac{1}{2}, 0, 0)$. If we increase the budget to $B = 27$, then again $k^* = 2$ because k^* does not depend on B. But $\Delta^* = 3\frac{6}{7}$, and since $\Delta < \Delta^*$, we have $\boldsymbol{u} = (2\frac{1}{2}, 2\frac{1}{2}, \frac{1}{2}, \frac{1}{2})$.

Sample solutions for eight stakeholders appear in Table 2.2. When Δ is 1 or 3, the utility allotments for the four more efficient stakeholders are Δ greater than those for the remaining stakeholders. When Δ is 4 or 6, some stakeholders receive utility Δ and others zero, with one stakeholder receiving an intermediate value. Figure 11.2 displays the distribution as Δ ranges from 0 to 15. The heavy black curve (average utility) indicates that the *marginal* cost of fairness is greater when the solution is closer to equality (smaller Δ). This contrasts with alpha fairness, where the marginal cost of fairness is greater when the solution is closer to utilitarianism.

Interestingly, the optimal solution $\boldsymbol{u}$ given in Proposition 11.2 satisfies $u_i - u_j \leq \Delta$ for all i and j. Thus there is always an optimal solution in the utilitarian region of the contour plot when one is maximizing subject to a budget constraint. In fact, this is true for any feasible set that is closed under domination, that is, any feasible set S such that $\boldsymbol{u}' \in S$ whenever

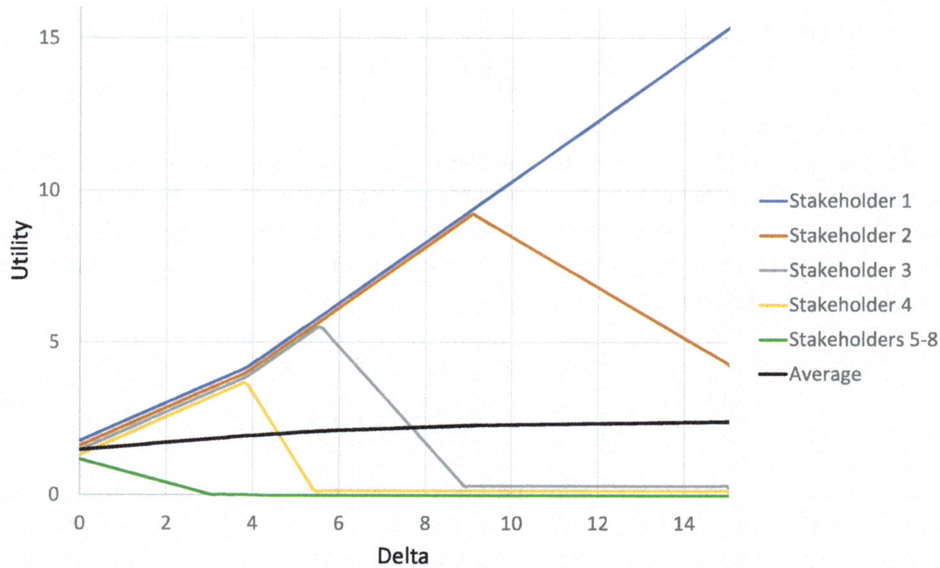

Fig. 11.2 Utility distribution over eight stakeholders using the equity threshold criterion. Lines that coincide are slightly offset to clarify the pattern

$\boldsymbol{u} \in S$ and $\boldsymbol{u}' \leq \boldsymbol{u}$. For example, the feasible set in Fig. 11.1 is closed under domination, and the optimal solution shown in fact lies in the utilitarian region. By contrast, the feasible set depicted in Fig. 11.3 is not closed under domination, and no solutions lie in the utilitarian region. However, such solutions can exist in other feasible sets that lack closure under domination, because closure is a sufficient but not a necessary condition for their existence.

Proposition 11.3 *If S is closed under domination, and some point is optimal in S for the equity threshold criterion, then some point in the utilitarian region of S is optimal.*

Proof We wish to show that if some $\bar{\boldsymbol{u}}$ maximizes $W_\Delta(\boldsymbol{u})$ over S, then some point $\boldsymbol{u}' \in S$ for which $\max_{ij}\{u'_i - u'_j\} \leq \Delta$ maximizes $W_\Delta(\boldsymbol{u})$ over S. Such a point $\boldsymbol{u}'$ is obtained by setting $u'_i = \min\{\bar{u}_i, \bar{u}_{\min} + \Delta\}$ for all i. We have $\boldsymbol{u}' \in S$ because $\boldsymbol{u}' \leq \bar{\boldsymbol{u}}$. It is also easily checked that $\max_{ij}\{u'_i - u'_j\} \leq \Delta$ and that $W_\Delta(\boldsymbol{u}') = W_\Delta(\bar{\boldsymbol{u}})$.

Finally, we will see in Chap. 13 that the optimal solution given in Proposition 11.2 maximizes total utility subject to a bound Δ on the range of utility values. This means that when the only constraint is a linear budget restriction, maximizing an equity threshold criterion is equivalent to maximizing total utility while treating the threshold as a bound on the utility range. The two models are not equivalent in general, however. As is evident in Fig. 11.3, the utilities can have a range greater than Δ in an equity threshold solution.

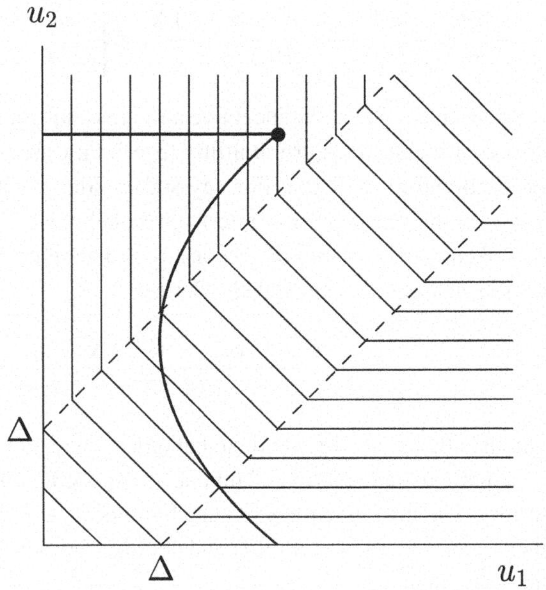

Fig. 11.3 Solution (black dot) of a two-person equity threshold problem for which no solutions lie in the utilitarian region of the feasible set

11.3 Hierarchical Distribution

The equity threshold SWF is not regionally decomposable. The counterexample exhibited in the previous section serves as a counterexample here. In this case, the unique optimal solution of the national problem is $(\bar{u}_1, \bar{u}_2, \bar{u}_3) = (1, 1, 0)$ and $(\bar{y}_1, \bar{y}_2) = (1, 0)$, while the only two optimal solutions of the region 1 subproblem are $(u_1, u_2) = (1.5, 0.5)$ and $(0.5, 1.5)$. Thus an optimal solution of the regional problem cannot be part of an optimal solution of the national problem.

11.4 Incentives and Sharing

Marginal analysis is very different for the regimes corresponding to $\Delta < \Delta^*$ and $\Delta \geq \Delta^*$. In the former case, an efficiency improvement by any stakeholder increases everyone's utility level by an equal absolute amount. When $\Delta \geq \Delta^*$, only one specific stakeholder's utility is affected by efficiency improvements anywhere in the population. Since raising the budget B can cause a transition from the latter case to the former, it can substantially boost the degree of benefit sharing as well as raise overall benefit levels.

We first suppose $\Delta < \Delta^*$. In this case, the rate of change is the same for every stakeholder with respect to every cost coefficient. Specifically, it is

$$\frac{\partial u_j^*}{\partial a_i} = -\frac{B + \big(a(n) - a(k^*)\big)\Delta}{a(n)^2} = \begin{cases} -u_j^*/a(n), & \text{for } j = 1, \ldots, k^* \\ -(u_j^* + \Delta)/a(n), & \text{for } j = k^* + 1, \ldots, n \end{cases}$$

for all i and j. Since the incentive and cross-subsidy rates are measured with respect to the cost coefficients and current utility levels, they are proportional to the former and inversely proportional to the latter. For any stakeholder j with current utility level of $u_0 + \Delta$ (i.e., $j \le k^*$), the cross-subsidy rate from stakeholder i is $a_i/a(n)$, which is the incentive rate of stakeholder j when $i = j$. For any stakeholder j with utility level Δ (i.e., $j > k^*$), the cross-subsidy rate from stakeholder i is

$$\frac{B + \big(a(n) - a(k^*)\big)\Delta}{B - a(k^*)\Delta} \cdot \frac{a_i}{a(n)} = \left(1 + \frac{\Delta}{u_j^*}\right)\frac{a_i}{a(n)}$$

which is the incentive rate of stakeholder j when $i = j$.

Thus a given reduction of anyone's cost coefficient benefits everyone equally (in absolute terms). The incentive rates of stakeholders $k^* + 1, \ldots, n$, as well as cross-subsidy rates to them, are higher than for other stakeholders because they start with a smaller utility level of u_0 rather than $u_0 + \Delta$.

In the above example, when $B = 27$ we have $\Delta < \Delta^*$, and the rate of change of any stakeholder's utility with respect to any cost coefficient is $-65/26^2 = -0.09615$. For stakeholder 1, a 1% reduction in the cost coefficient yields approximately a 0.115% utility increase for stakeholders 1 and 2 and a 0.577% utility increase for stakeholders 3 and 4. For stakeholder 2, a 1% cost reduction yields a 0.154% utility increase for stakeholders 1 and 2 and 0.769% increase for stakeholders 3 and 4. The results are similar when stakeholders 3 and 4 reduce cost.

We now suppose $\Delta \ge \Delta^*$. In this case, only stakeholder k is affected by marginal changes in cost coefficients, because the other stakeholders have constant utility at the margin, equal to Δ or 0. Stakeholder k is therefore the only one with a positive incentive rate. Since it has no cross subsidies with other stakeholders, its incentive rate is 1, which can be verified as follows:

$$-\frac{\partial u_k^*}{\partial a_k} \cdot \frac{a_k}{u_k^*} = \frac{B - a(k-1)\Delta}{a_k^2} \cdot \frac{a_k}{u_k^*} = 1$$

The cross-subsidy rate from a stakeholder $i > k$ to stakeholder k is

$$-\frac{\partial u_k^*}{\partial a_i} \cdot \frac{a_i}{u_k^*} = \frac{\Delta}{a_k} \cdot \frac{a_i}{u_k^*} = \frac{a_i \Delta}{B - a(k-1)\Delta}$$

and the cross-subsidy rates with other stakeholders are zero. Thus stakeholder k benefits proportionately from its cost reductions, while other stakeholders do not benefit from anyone's cost reduction (including their own) at the margin. The surplus created by a cost reduction by any stakeholder $i \in \{1, \ldots, k-1\}$ is absorbed completely by stakeholder k, in an amount roughly proportional to stakeholder i's current cost.

In the above example, when $B = 12$ we have $\Delta \geq \Delta^*$, $k^* = 2$ and $k = 2$, which means stakeholder 2 receives the benefit of cost reductions. It receives a 1% increase in its utility allotment for each 1% reduction in it own cost, and as it happens, for each 1% reduction in stakeholder 1's cost.

When $\Delta < \Delta^*$, the technology payoff gradient is a positive scalar multiple of

$$(u_0 + \Delta, \ldots, u_0 + \Delta, u_0, \ldots, u_0)$$

where there are k^* components $u_0 + \Delta$. Thus technological investment should be directed toward all stakeholders, with somewhat more emphasis on the more efficient ones. When $\Delta \geq \Delta^*$, the gradient is

$$\big(\Delta, \ldots, \Delta,\ B - a(k-1)\Delta,\ 0, \ldots, 0\big)$$

which begins with $k - 1$ Δ's. The less efficient stakeholders should receive no investment in this regime, as one might expect because the SWF is closer to utilitarianism.

Reference

1. Williams, A., Cookson, R.: Equity in Health. Culyer, A.J. and Newhouse, J.P. editors, Handbook of Health Economics (2000)

Utility Threshold Criterion with Leximax Fairness 12

12.1 Mathematical Formulation

A utility threshold function in which a maximin criterion represents equity inherits a weakness of the maximin criterion: it ignores all but the very lowest utility value. This can result in solutions that are insensitive to the plight of disadvantaged stakeholders other than the very worst off. This issue is addressed in [1] by combining utilitarianism with a leximax rather than a maximin criterion. An added benefit of this approach is that it generally avoids purely utilitarian and purely leximax solutions, even when there is a single budget constraint with no upper bounds on utilities.

However, to implement this fairness criterion it is necessary to solve a sequence of optimization problems $P_1, \ldots, P_n$ rather than a single problem. Each problem P_k determines the value $\bar{u}_{i_k}$ of the kth smallest utility in the socially optimal solution $\bar{\boldsymbol{u}}$. Problem P_k accomplishes this by maximizing the social welfare function W_k, where W_1 is the utility threshold SWF described in Chap. 10, and the remaining SWFs are

$$W_k\big(\boldsymbol{u}(I_k)\big) = (n-k+1)u_{\min} + \sum_{i\in I_k}\big(u_i - \bar{u}_{i_1} - \Delta\big)^+$$

where $(\alpha)^+ = \max\{\alpha, 0\}$, $u_{\min} = \min_{i\in I_k}\{u_i\}$, and $\bar{u}_{i_1}, \ldots \bar{u}_{i_{k-1}}$ are the socially optimal utility values determined by problems $P_1, \ldots, P_{k-1}$. Also $I_k = N \setminus \{i_1, \ldots, i_{k-1}\}$ indexes the utility variables u_i yet to be determined, and $W(\boldsymbol{u}(I_k))$ is a function of these variables only.

Ö. Elçi et al., *The Structure of Fair Solutions*, Synthesis Lectures on Operations Research and Applications, https://doi.org/10.1007/978-3-031-82190-5_12

The first problem P_1 simply maximizes $W_1(\boldsymbol{u})$ subject to $(\boldsymbol{u}, \boldsymbol{x}) \in S$. The remaining problems P_k for $k = 2, \ldots, n$ are

$$\max_{\boldsymbol{u},\boldsymbol{x}} \left\{ W_k\big(\boldsymbol{u}(I_k)\big) \;\middle|\; \begin{array}{l} u_i \geq \bar{u}_{i_{k-1}},\ i \in I_k \\ u_{i_j} = \bar{u}_{i_j},\ j = 1, \ldots, k-1 \\ (\boldsymbol{u}, \boldsymbol{x}) \in S \end{array} \right\} \tag{12.1}$$

Actually, one need only solve $P_1, \ldots, P_{k^*}$ (rather than $P_1, \ldots, P_n$) where k^* is the smallest k for which $\bar{u}_{i_k} - \bar{u}_{i_1} \geq \Delta$. At this point, all the remaining utilities (i.e., all u_i for $i \in I_{k^*}$) are in the utilitarian region, and one can set $\bar{u}_i = \tilde{u}_i$ for $i \in I_{k^*}$, where $\tilde{u}_i$ is the optimal value of u_i in the solution of P_{k^*}.

MILP models of social welfare functions $W_k(\boldsymbol{u}(I_k))$ are given in [1]. Problem P_1 has the model (10.2). Problem P_k for $k = 2, \ldots, n$ can be written as follows. As in the case of model (10.2), it is necessary to make the harmless assumption that there is an M such that $u_i - u_j \leq M$ for i, j to ensure MILP representability.

$$\begin{aligned}
&\max\ z \\
&z \leq (n-k+1)\sigma + \sum_{i \in I_k} v_i \\
&0 \leq v_i \leq M\delta_i,\ \ i \in I_k \\
&v_i \leq u_i - \bar{u}_{i_1} - \Delta + M(1-\delta_i),\ \ i \in I_k \\
&\sigma \leq \bar{u}_{i_1} + \Delta,\ \ \sigma \leq w \\
&w \leq u_i \leq w + M(1-\epsilon_i),\ i \in I_k \\
&\sum_{i \in I_k} \epsilon_i = 1 \\
&w \geq \bar{u}_{i_{k-1}} \\
&u_i - \bar{u}_{i_1} \leq M,\ \ \delta_i, \epsilon_i \in \{0, 1\},\ i \in I_k \\
&(\boldsymbol{u}, \boldsymbol{x}) \in S
\end{aligned}$$

where z, w, σ, v_i, δ_i and ϵ_i are auxiliary variables. The validity of this model is proved in [1], and its practicality demonstrated by solving realistic problem instances in a few seconds.

12.2 Socially Optimal Distribution

We showed earlier (Proposition 10.2) that when the social welfare problem is solved subject only to a budget constraint, the solution of P_1 is purely utilitarian or purely maximin. The socially optimal solution $\bar{\boldsymbol{u}}$ that results from solving the sequence of problems $P_1, \ldots, P_n$ is, in general, neither utilitarian nor leximax, as illustrated in Table 2.2. It is convenient to suppose on this occasion that $a_n \leq \cdots \leq a_1$, rather than the reverse, because this allows problems $P_1, \ldots, P_n$ to determine the solution values of $u_1, \ldots, u_n$ in that order. The resulting solution is as follows.

Proposition 12.1 *Suppose $a_n \le \cdots \le a_1$. Let m be the smallest index k for which the following is violated:*

$$\Delta \le B\left(\frac{1}{a_n} - \frac{n-k+1}{a(k:n)}\right) \tag{12.2}$$

where $a(k:n) = \sum_{j=k}^{n} a_j$. We regard (12.2) as violated if $k = n$. Then a socially optimal solution $\bar{\boldsymbol{u}}$ of the leximax threshold problem subject to a budget constraint is given by the following:

$$\bar{u}_i = \begin{cases} 0, & \text{for } i = 1, \ldots, m-1 \\ B/a(m:n), & \text{for } i = m, \ldots, n \end{cases} \tag{12.3}$$

To prove Proposition 12.1, we note that when the social welfare problem has only a budget constraint, problem P_k has the form:

$$\min_{\boldsymbol{u}(I_k)} \left\{ W_k\big(\boldsymbol{u}(I_k)\big) \;\middle|\; \begin{array}{c} u_i \ge \bar{u}_{i_{k-1}},\ i \in I_k \\ \sum_{i \in I_k} a_i u_i \le B_k \end{array} \right\} \tag{12.4}$$

where

$$W_k\big(\boldsymbol{u}(I_k)\big) = (n-k+1)u_{\min} + \sum_{i \in I_k} (u_i - \bar{u}_{i_1} - \Delta)^+$$

and $u_{\min} = \min_{i \in I_k}\{u_i\}$. To simplify notation, we offset the utilities with the change of variable $u_i' = u_i - \bar{u}_{i_{k-1}}$. Problem P_k now becomes a problem P_k' of the form:

$$\max_{\boldsymbol{u}'(I_k)} \left\{ W_k'\big(\boldsymbol{u}'(I_k)\big) \;\middle|\; \begin{array}{c} u_i' \ge 0,\ i \in I_k \\ \sum_{i \in I_k} a_i u_i' \le B_k' \end{array} \right\} \tag{12.5}$$

where $B_k' = B_k - a(I_k)\bar{u}_{i_{k-1}}$ with $a(I_k) = \sum_{i \in I_k} a_i$, and

$$W_k'\big(\boldsymbol{u}'(I_k)\big) = (n-k+1)u_{\min}' + \sum_{i \in I_k} (u_i' - \Delta')^+$$

with $\Delta' = \Delta - (\bar{u}_{i_{k-1}} - \bar{u}_{i_1})$. It is useful to reformulate (12.5) as follows:

$$\max_{v_0, \boldsymbol{v}} \left\{ \overline{W}_k(v_0, \boldsymbol{v}) \;\middle|\; \begin{array}{c} v_0 \ge 0,\ v_i \ge 0,\ i \in I_k \\ a(I_k)v_0 + \sum_{i \in I_k} a_i v_i \le B_k' \end{array} \right\} \tag{12.6}$$

where

$$\overline{W}(v_0, \boldsymbol{v}) = (n - k + 1)v_0 + \sum_{i \in I_k} (v_i + v_0 - \Delta')^+$$

Lemma 12.1 *Formulation* (12.6) *has the same optimal value as* (12.5)*, and solution* $\boldsymbol{u}'(I_k)$ *is optimal in* (12.5) *if and only if* $(v_0, \boldsymbol{v})$ *is optimal in* (12.6)*, where* $v_0 = u'_{\min}$ *and* $v_i = u'_i - u'_{\min}$ *for* $i \in I_k$.

Proof It suffices to show that (a) if $\boldsymbol{u}'(I_k)$ is optimal in (12.5), then $(v_0, \boldsymbol{v})$ is optimal in (12.6), where $v_0 = u'_{\min}$ and $v_i = u'_i - u'_{\min}$, and (b) if $(v_0, \boldsymbol{v})$ is optimal in (12.6), then $\boldsymbol{u}'(I_k)$ is optimal in (12.5), where $u'_i = v_0 + v_i$ for $i \in I_k$. To show (a), suppose that $\boldsymbol{u}'(I_k)$ is optimal in (12.5). Then $(v_0, \boldsymbol{v})$ is feasible in (12.6), given the constraints of (12.5). Also $\overline{W}_k(v_0, \boldsymbol{v}) = W'_k(\boldsymbol{u}'(I_k))$, and so $(v_0, \boldsymbol{v})$ is optimal in (12.6).

To show (b), suppose that $(v_0, \boldsymbol{v})$ is optimal in (12.6). Set $u'_i = v_0 + v_i$ for $i \in I_k$, which implies $u'_{\min} = v_0 + v_{\min}$, where $v_{\min} = \min_{i \in I_k}\{v_i\}$. It is easily checked that $\boldsymbol{u}'(I_k)$ is feasible in (12.5). Then $W'_k(\boldsymbol{u}'(I_k))$ becomes

$$(n - k + 1)(v_{\min} + v_0) + \sum_{i \in I_k} (v_i + v_0 - \Delta')^+$$

This is no smaller than $\overline{W}(v_0, \boldsymbol{v})$ because $v_{\min} \geq 0$. Since $\overline{W}(v_0, \boldsymbol{v})$ is the optimal value of (12.6), $\boldsymbol{u}'(I_k)$ is optimal in (12.5). □

We next show that (12.5) has a utilitarian solution when the following holds, and a maximin solution otherwise:

$$\Delta' \leq B'_k \left(\frac{1}{a_n} - \frac{n - k + 1}{a(I_k)} \right) \tag{12.7}$$

where we regard (12.7) as false when $k = n$.

Lemma 12.2 *If* $B'_k \geq 0$*, problem* (12.5) *has a optimal solution* $\boldsymbol{u}'(I_k)$ *in which*

$$\boldsymbol{u}'(I_k) = \begin{cases} \dfrac{B'_k}{a_n} \boldsymbol{e}_n, & \textit{if (12.7) holds} \\ \dfrac{B'_k}{a(I_k)} \boldsymbol{e}(I_k), & \textit{otherwise} \end{cases}$$

where $\boldsymbol{e}_n = (0, \ldots, 0, 1)$ *and* $\boldsymbol{e}(I_k)$ *is a vector of* $|I_k|$ *ones.*

Proof Due to Lemma 12.1, it suffices to show that problem (12.6) has an optimal solution in which

$$(v_0, \boldsymbol{v}) = \begin{cases} \big(0, \dfrac{B'_k}{a_n}\boldsymbol{e}_n\big), & \text{if (12.7) holds} \\ \big(\dfrac{B'_k}{a(I_k)}, \boldsymbol{0}\big), & \text{otherwise} \end{cases}$$

Since we solve P_k only for $k \le k^*$, we know that $\bar{u}_{i_k} - \bar{u}_{i_1} \le \Delta$, which implies $\Delta' \ge 0$. Thus we have

$$\overline{W}_k\big(0, \frac{B'_k}{a_n}\boldsymbol{e}_n\big) = \left(\frac{B'_k}{a_n} - \Delta'\right)^+ \tag{12.8}$$

$$\overline{W}_k\big(\frac{B'_k}{a(I_k)}, \boldsymbol{0}\big) = (n-k+1)\frac{B'_k}{a(I_k)} + (n-k+1)\left(\frac{B'_k}{a(I_k)} - \Delta'\right)^+ \tag{12.9}$$

We will show that $(v_0, \boldsymbol{v}) = (0, (B'_k/a_n)\boldsymbol{e}_n)$ is an optimal solution of (12.6) if (12.7) holds, and $(v_0, \boldsymbol{v}) = (B'_k/a(I_k), \boldsymbol{0})$ is an optimal solution otherwise. We first observe that the objective function $\overline{W}_k(v_0, \boldsymbol{v})$ is convex. It follows that some extreme point of the feasible set of (12.6) is optimal. Yet every extreme point is the solution of some linearly independent subset T of $n+1$ of the following equations:

$$\begin{aligned} & a(I_k)v_0 + \sum_{i \in I_k} v_i = B'_k && (a) \\ & v_0 = 0 && (b) \\ & v_i = 0, \; i \in I_k && (c) \end{aligned}$$

We can suppose T contains (a), since otherwise the corresponding extreme point $(v_0, \boldsymbol{v}) = (0, \boldsymbol{0})$ is clearly dominated by $(0, (B'_k/a_n)\boldsymbol{e}_n)$ and $(B'_k/a(I_k), \boldsymbol{0})$. Then T either contains (b) and all but one equation $v_j = 0$ in (c), or else all equations in (c). The former yields extreme point $(v_0, \boldsymbol{v}) = (0, (B'_k/a_j)\boldsymbol{e}_j)$ and the latter $(v_0, \boldsymbol{v}) = (B'_k/a(I_k), \boldsymbol{0})$.

We first suppose that (12.7) holds and show that $(v_0, \boldsymbol{v}) = (0, (B'_k/a_n)\boldsymbol{e}_n)$ is optimal. For this, it suffices to show

$$\overline{W}_k\big(0, \frac{B'_k}{a_j}\boldsymbol{e}_j\big) \le \overline{W}_k\big(0, \frac{B'_k}{a_n}\boldsymbol{e}_n\big), \text{ all } j \in I_k \tag{12.10}$$

$$\overline{W}_k\big(\frac{B'_k}{a(I_k)}, \boldsymbol{0}\big) \le \overline{W}_k\big(0, \frac{B'_k}{a_n}\boldsymbol{e}_n\big) \tag{12.11}$$

Due to (12.8)–(12.9), these are, respectively, equivalent to

$$\left(\frac{B'_k}{a_j} - \Delta'\right)^+ \leq \left(\frac{B'_k}{a_n} - \Delta'\right)^+, \text{ all } j \in I_k \tag{12.12}$$

$$(n-k+1)\frac{B'_k}{a(I_k)} + (n-k+1)\left(\frac{B'_k}{a(I_k)} - \Delta'\right)^+ \leq \left(\frac{B'_k}{a_n} - \Delta'\right)^+ \tag{12.13}$$

But (12.12) holds simply because $a_n \leq a_j$ for all $j \in I_k$. To show (12.13), we distinguish three cases. *Case 1:* $\Delta' \leq B'_k/a(I_k)$. In this case, to establish (12.13) it suffices to show that

$$(n-k+1)\frac{B'_k}{a(I_k)} + (n-k+1)\left(\frac{B'_k}{a(I_k)} - \Delta'\right) \leq \frac{B'_k}{a_n} - \Delta'$$

This can be written as

$$\Delta' \geq \frac{B'_k}{n-k}\left(\frac{n-k+1}{a(I_k)} - \frac{1}{a_n}\right)$$

But since $a(I_k) \geq (n-k+1)a_n$, to show this it suffices to show

$$\frac{n-k}{a_n} \geq \frac{2}{(n-k)a_n}, \quad \text{or } n-k \geq \frac{2}{n-k}$$

This holds if $n-k \geq 2$. But we can suppose $\Delta' > 0$, since otherwise the problem is purely utilitarian and therefore has a purely utilitarian solution. This and (12.7) imply $n-k \geq 2$. *Case 2:* $B'_k/a(I_k) < \Delta' \leq B_k/a_n$. Here it suffices to show that

$$(n-k+1)\frac{B'_k}{a(I_k)} \leq \frac{B'_k}{a_n} - \Delta'$$

But this is equivalent to (12.7). *Case 3:* $\Delta' > B'_k/a_{\ell_k}$. This case cannot occur because it violates (12.7). Since case 1 or 2 must obtain, (12.13) follows.

We now suppose that (12.7) is violated, so that

$$\Delta' > B'_k\left(\frac{1}{a_n} - \frac{n-k+1}{a(I_k)}\right) \tag{12.14}$$

We wish to show that $(v_0, \boldsymbol{v}) = (B'_k/a(I_k), \mathbf{0})$ is optimal. For this it suffices to show

$$\overline{W}_k\left(0, \frac{B'_k}{a_j}\boldsymbol{e}_j\right) \leq \overline{W}_k\left(\frac{B'_k}{a(I_k)}, \mathbf{0}\right), \text{ all } j \in I_k \tag{12.15}$$

which is equivalent to

$$\left(\frac{B'_k}{a_j} - \Delta'\right)^+ \leq (n-k+1)\frac{B'_k}{a(I_k)} + (n-k+1)\left(\frac{B'_k}{a(I_k)} - \Delta'\right)^+, \text{ all } j \in I_k \tag{12.16}$$

We again consider three cases.

Case 1: $\Delta' \le B'_k/a(I_k)$. Given this, it suffices to show the following for any $j \in I_k$:

$$\frac{B'_k}{a_j} - \Delta' \le (n-k+1)\frac{B'_k}{a(I_k)} + (n-k+1)\Big(\frac{B'_k}{a(I_k)} - \Delta'\Big)$$

which can be written as

$$(n-k)\Delta' \le B'_k\Big(2\frac{n-k+1}{a(I_k)} - \frac{1}{a_j}\Big)$$

Due to the case hypothesis, it suffices to show

$$B'_k\frac{n-k}{a(I_k)} \le B'_k\Big(2\frac{n-k+1}{a(I_k)} - \frac{1}{a_j}\Big), \quad \text{or} \quad \frac{1}{a_j} \le \frac{n-k+2}{a(I_k)}$$

Since $a_j \ge a_n$ and $a(I_k) \ge (n-k+1)a_n$, for this it suffices to show

$$\frac{1}{a_n} \le \frac{n-k+2}{n-k+1}\cdot\frac{1}{a_n}$$

which obviously holds. *Case 2:* $B'_k/a(I_k) < \Delta' \le B'_k/a_j$. In this case, it suffices to show

$$\frac{B'_k}{a_j} - \Delta' \le (n-k+1)\frac{B'_k}{a(I_k)}, \quad \text{or} \quad \Delta' \ge B_k\left(\frac{1}{a_j} - \frac{n-k+1}{a(I_k)}\right), \text{ all } j \in I_k$$

This follows from (12.14) because $a_j \ge a_n$. *Case 3.* $\Delta' > B'_k/a_n$. Since $a_n \le a_j$ for any $j \in I_k$, it suffices in this case to show that

$$0 \le (n-k+1)\frac{B'_k}{a(I_k)}$$

But this is true because we are given that $B'_k \ge 0$. Since one of the three cases must obtain, the proof is complete.

Proof of Proposition 12.1. Let $i_j = j$ for $j = 1, \ldots, n$. It suffices to prove that the following claims are true in some socially optimal solution $\bar{\boldsymbol{u}}$.

$$\bar{u}_k = 0 \text{ for } k = 1, \ldots, m-1 \tag{12.17}$$

$$\bar{u}_k = B/a(m\!:\!n) \text{ for } k = m, \ldots, n \tag{12.18}$$

Proof of (12.17). The argument is by induction. For $k = 1$, we know that (10.7) is satisfied, by definition of m. Thus by Proposition 10.2, an optimal solution of P_1 is $\tilde{\boldsymbol{u}} = (0, \ldots, 0, B/a(n))$, and we can set $\bar{u}_1 = \min\{\tilde{u}_i \mid i = 1, \ldots, n\} = 0$. We now suppose that (12.17) is true for $k = 1, \ldots, \ell - 1$ and show it is also true for $k = \ell$ in some optimal solution. We note first that $I_\ell = \{1, \ldots, \ell - 1\}$. Then since $\bar{u}_i = 0$ for $i = 1, \ldots, \ell - 1$, we have that $B_\ell = B$ and $B'_\ell = B_\ell - a_{\ell-1}\bar{u}_{\ell-1} = B_\ell = B$. Also $\Delta' = \Delta - (\bar{u}_{\ell-1} - \bar{u}_1) = \Delta$. Thus (12.7) reduces to (12.2). But (12.2), and therefore (12.7), are satisfied by the definition of m. Since $B'_\ell \geq 0$, Lemma 12.2 implies that $\tilde{\boldsymbol{u}}' = (0, \ldots, 0, B'_\ell/a_n) = (0, \ldots, 0, B/a_n)$ solves P'_ℓ. This implies that $\tilde{u} = \tilde{u}' - \bar{u}_{\ell-1}\boldsymbol{e} = (0, \ldots, 0, B/a_n)$ solves P_ℓ, and we can let $\bar{u}_\ell = 0$.

Proof of (12.18). The argument is again by induction. First suppose (12.17) is true and consider $k = m$ in (12.18). We have $B'_m = B$ and $\Delta' = \Delta$ for the same reason as above so that (12.2) again reduces to (12.7). But (12.2), and therefore (12.7), is violated, by definition of m. Due to this and $B'_m \geq 0$, Lemma 12.2 implies that P'_m has a solution $\tilde{\boldsymbol{u}}' = (B/a_M, \ldots, B/a_M)$, which implies that P_m has a solution $\tilde{\boldsymbol{u}} = \tilde{\boldsymbol{u}}'_M - \bar{u}_{m-1}\boldsymbol{e} = (B/a_M, \ldots, B/a_M)$ since $\bar{u}_{m-1} = 0$. We can therefore set $\bar{u}_m = B/a_M$. We now suppose (12.17) is true, and (12.18) is true for $k = m, \ldots, \ell - 1$, and we show (12.18) is also true for $k = \ell$ in some optimal solution. By the induction hypothesis,

$$B_\ell = B - \sum_{i=m}^{\ell-1} a_i \bar{u}_m = B - \frac{B}{a(m:n)} \sum_{i=m}^{\ell-1} a_i$$

and

$$B'_\ell = B_\ell = \bar{u}_{\ell-1} \sum_{i=\ell}^{n} a_i = B - \frac{B}{a(m:n)} \left(\sum_{i=m}^{\ell-1} a_i + \sum_{i=\ell}^{n} a_i \right) = B - \frac{B}{a(m:n)} \cdot a(m:n) = 0$$

Since $B'_\ell = 0$, the only feasible solution of P'_ℓ is $(\tilde{u}_\ell, \ldots, \tilde{u}_n) = (0, \ldots, 0)$. We can therefore set $\bar{u}'_\ell = 0$, which means $\bar{u}_\ell = B/a(m:n)$, as claimed. □

Thus, several of the more efficient stakeholders receive equal shares of utility, with the remaining stakeholders receiving nothing. Utility can be spread more broadly by increasing Δ, even to a point where everyone receives an equal share. A utility threshold criterion that combines utilitarian and leximax elements can therefore yield a variety of solutions, even when there is only a single budget constraint. The variety is still greater when there are lower and/or upper bounds on utilities.

As an example, suppose we wish to maximize the utility threshold function with leximax, subject to the budget constraint $6u_1 + 4u_2 + 3u_3 + 2u_4 \leq 30$. We obtain the results for various values of Δ that are displayed in Table 12.1. The solution is utilitarian for $\Delta \leq 3$, maximin for $\Delta > 11$, and neither for intermediate values of Δ.

Solutions for eight stakeholders and three values of Δ appear in Table 2.2, where $\boldsymbol{a} = (5, 6 \ldots, 12)$ and $B = 100$. Figure 12.1 contains a plot of the distributions corresponding to values of Δ ranging from 0 to 10. The utility threshold SWF with leximax is the only fairness criterion we study for which the utility curves represent step functions.

Table 12.1 Solutions of a utility threshold problem with leximax

Δ range	m	$(u_1, \ldots, u_4)$
$0 \le \Delta \le 3$	4	$(0, 0, 0, 15)$
$3 < \Delta \le 5$	3	$(0, 0, 6, 6)$
$5 < \Delta \le 11$	2	$(0, 3\frac{1}{3}, 3\frac{1}{3}, 3\frac{1}{3})$
$11 < \Delta < \infty$	1	$(2, 2, 2, 2)$

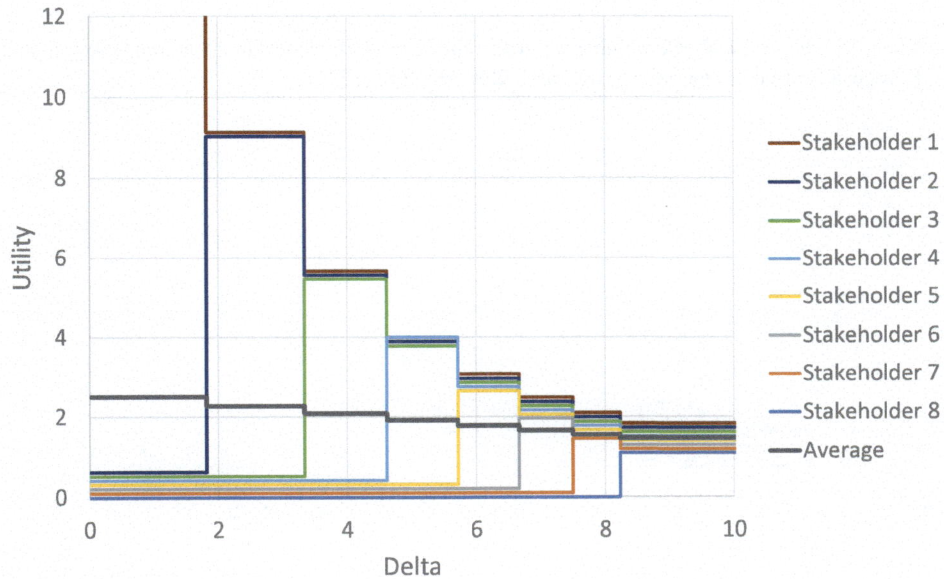

Fig. 12.1 Utility distribution over eight stakeholders using the utility threshold criterion with leximax. Lines that coincide are slightly offset vertically to clarify the pattern

12.3 Hierarchical Distribution

The counterexample exhibited in the two previous chapters shows a lack of regional decomposability.

12.4 Incentives and Sharing

The incentive rate for stakeholder i is $a_i/a(m:n)$ if $i \ge m$ and zero otherwise. This means that less efficient stakeholders who are allotted utility have a proportionately greater incentive to improve efficiency. The cross-subsidy rate from stakeholder i to stakeholder j is likewise $a_i/a(m:n)$ if $i, j \ge m$, and otherwise zero. This means there is sharing rather than competition, although this effect exists at the margin only among those who already benefit from the optimal distribution. The technology payoff gradient is a positive scalar multiple of $(0, \ldots, 0, a_m, a_{m+1}, \ldots, a_n)$, which recommends technology investments in the $n - m + 1$ more efficient stakeholders so as to increase their efficiencies by a percentage inversely proportional to their current efficiency.

Reference

1. Chen, V., Hooker, J.N.: Combining leximax fairness and efficiency in an optimization model. European Journal of Operational Research **299**, 235–248 (2022)

Inequality Bounds: Range

13

13.1 Mathematical Formulation

In this and the next chapter, we explore distributions that maximize total utility subject to an upper bound on some inequality measure. In the present chapter, we examine a bound on the *range* $\max_{ij} |u_i - u_j|$ of utilities. That is, we solve

$$\max_{\boldsymbol{u},\boldsymbol{x}} \left\{ \sum_i u_i \;\middle|\; u_i - u_j \leqslant R, \text{ all } i, j; \;\; (\boldsymbol{u}, \boldsymbol{x}) \in S \right\} \tag{13.1}$$

where $R \geq 0$ is the desired upper bound. The absolute difference $|u_i - u_j|$ is replaced by $u_i - u_j$ to achieve linearity without affecting the solution.

One might consider maximizing utility subject to the relative range rather than the absolute range. The relative range is normalized by the mean utility and is therefore given by $\max_{ij}\{u_i - u_j\}/\bar{u}$. However, the resulting model is not useful for obtaining fair solutions. To see this, note that maximizing the total utility is equivalent to maximizing the mean utility. The problem that maximizes mean utility subject to an upper bound R on range is

$$\max_{\boldsymbol{u},\boldsymbol{x}} \left\{ \bar{u} \;\middle|\; \frac{u_i - u_j}{R} \leqslant \bar{u}, \text{ all } i, j; \;\; (\boldsymbol{u}, \boldsymbol{x}) \in S \right\} \tag{13.2}$$

The range constraints place *lower bounds* on an objective function that is maximized. Suppose we solve the problem without the range constraints to obtain a purely utilitarian solution $\boldsymbol{u}^*$ with mean utility $\bar{u}^*$. Then any optimal solution of (13.2) with the range constraints, if such a solution exists, must have mean utility $\bar{u}^*$. The model, thus, can provide a solution only if it is purely utilitarian. We therefore maximize utility subject to a bound on the absolute utility range, as in (13.1).

Ö. Elçi et al., *The Structure of Fair Solutions*, Synthesis Lectures on Operations Research and Applications, https://doi.org/10.1007/978-3-031-82190-5_13

13.2 Socially Optimal Distribution

As in previous chapters, we analyze the socially optimal solution subject to a linear budget constraint, augmented in the present case by a bound R on the utility range. The problem is therefore

$$\max_{\boldsymbol{u}} \left\{ \sum_j u_j \;\middle|\; u_i - u_j \leqslant R, \text{ all } i, j; \;\; \boldsymbol{a}^{\mathsf{T}}\boldsymbol{u} \leqslant B, \; \boldsymbol{u} \geq \boldsymbol{0} \right\} \tag{13.3}$$

We first note that the range-constrained problem is closely related to the equity threshold model of Chap. 11. The two models yield the same solution when solved subject only to a linear budget constraint, provided Δ in the threshold model is set equal to the maximum range R. The two fairness models are not equivalent in general, because as noted in Chap. 11, the threshold model can yield solutions with a range greater than Δ.

To see the equivalence of the models when solved with a linear budget constraint, recall from Proposition 11.2 that the equity threshold model has the linear formulation:

$$\max_{\boldsymbol{u}} \left\{ n\Delta + \sum_j v_j \;\middle|\; \begin{array}{l} w \leqslant u_j, \; v_j \leqslant u_j - \Delta, \; v_j \leqslant w, \text{ all } j \\ w \geq 0, \; \boldsymbol{a}^{\mathsf{T}}\boldsymbol{u} \leqslant B, \; \boldsymbol{u} \geq \boldsymbol{0} \end{array} \right\} \tag{13.4}$$

Lemma 13.1 *The models* (13.3) *and* (13.4) *are equivalent in the sense that when* $\Delta = R$*, they have the same optimal value, and there exists an optimal solution* $(\boldsymbol{u}, \boldsymbol{v}, w)$ *of* (13.4) *for which* $\boldsymbol{u}$ *is optimal in* (13.3).

Proof Let $f(\boldsymbol{u}, \boldsymbol{v}, w)$ be the objective function value of (13.4) for a given solution $(\boldsymbol{u}, \boldsymbol{v}, w)$, and let $g(\boldsymbol{u})$ be the objective function value of (13.3). For a given $\boldsymbol{u}$, define a solution $(\boldsymbol{u}, \boldsymbol{v}^{\boldsymbol{u}}, w^{\boldsymbol{u}})$ given by $w^{\boldsymbol{u}} = \min_j\{u_j\}$ and $v_j^{\boldsymbol{u}} = u_j - \Delta$ for all j. Then it is easily checked that $f(\boldsymbol{u}, \boldsymbol{v}^{\boldsymbol{u}}, w^{\boldsymbol{u}}) = g(\boldsymbol{u})$, and for any $\boldsymbol{u}$ feasible in (13.3), $(\boldsymbol{u}, \boldsymbol{v}^{\boldsymbol{u}}, w^{\boldsymbol{u}})$ is feasible in (13.4). Now from Proposition 11.2, there is an optimal solution $(\boldsymbol{u}, \boldsymbol{v}, w)$ of (13.4) such that $u_i - u_j \leqslant \Delta$ for all i, j, which means $\boldsymbol{u}$ is feasible in (13.3). It suffices to show that $\boldsymbol{u}$ is optimal in (13.3) and $f(\boldsymbol{u}, \boldsymbol{v}, w) = g(\boldsymbol{u})$. If $\boldsymbol{u}$ is not optimal in (13.3), then there is a feasible $\bar{\boldsymbol{u}}$ for which $g(\bar{\boldsymbol{u}}) > g(\boldsymbol{u})$. So we have

$$f(\bar{\boldsymbol{u}}, \boldsymbol{v}^{\bar{\boldsymbol{u}}}, w^{\bar{\boldsymbol{u}}}) = g(\bar{\boldsymbol{u}}) > g(\boldsymbol{u}) = f(\boldsymbol{u}, \boldsymbol{v}, w)$$

which contradicts the fact that $(\boldsymbol{u}, \boldsymbol{v}, w)$ is optimal in (13.4). Also

$$g(\boldsymbol{u}) = f(\boldsymbol{u}, \boldsymbol{v}^{\boldsymbol{u}}, w^{\boldsymbol{u}}) \leqslant f(\boldsymbol{u}, \boldsymbol{v}, w) \leqslant g(\boldsymbol{u})$$

where the first inequality is due to the optimality of $(\boldsymbol{u}, \boldsymbol{v}, w)$ and the second to the constraints $v_j \leqslant u_j - \Delta$ in (13.4). Thus $f(\boldsymbol{u}, \boldsymbol{v}, w) = g(\boldsymbol{u})$.

An optimal solution of the range-constrained problem now follows directly from Proposition 11.2 and Lemma 13.1.

Proposition 13.1 *Suppose* $a_1 \leqslant \cdots \leqslant a_n$, *and let*

$$k^* = \max\left\{k \;\middle|\; \frac{1}{a_k} > \frac{n}{a(n)}\right\}, \quad R^* = \frac{B}{a(k^*)}$$

where by convention $a(0) = 0$. *Then a utility distribution* $\boldsymbol{u}$ *that maximizes total utility subject to a budget constraint and a maximum range of* $R \geq 0$ *is given as follows. If* $R < R^*$, *then*

$$u_i = \begin{cases} u_0 + R, & \text{for } i = 1, \ldots, k^* \\ u_0, & \text{for } i = k^* + 1, \ldots, n \end{cases} \quad \text{where } u_0 = \frac{1}{a(n)}\Big(B - Ra(k^*)\Big) \tag{13.5}$$

If $R \geq R^*$, *then an optimal solution is*

$$u_i = \begin{cases} R, & \text{for } i = 1, \ldots, k-1 \\ \dfrac{1}{a(k)}\Big(B - Ra(k-1)\Big), & \text{for } i = k \\ 0, & \text{for } i = k+1, \ldots, n \end{cases} \tag{13.6}$$

where $k \leqslant k^*$ *and* k *satisfies*

$$\frac{B}{a(k)} < R \leqslant \frac{B}{a(k-1)}$$

We can illustrate this result with the example used in Chap. 11. Suppose we wish to maximize total utility $u_1 + \cdots + u_4$ subject to $3u_1 + 4u_2 + 7u_3 + 12u_4 \leqslant B$ with $B = 12$ and a maximum range $R = 2$. Then $k^* = 2$ and $R^* = 1\frac{5}{7}$. Since $R \geq R^*$ and $k = 2$, we have $\boldsymbol{u} = (2, 1\frac{1}{2}, 0, 0)$. If we increase the budget to $B = 27$, then again $k^* = 2$ because k^* does not depend on B. But $R^* = 3\frac{6}{7}$, and since $R < R^*$, we have $\boldsymbol{u} = (2\frac{1}{2}, 2\frac{1}{2}, \frac{1}{2}, \frac{1}{2})$.

Table 2.2 shows optimal distributions for $R = (1, 3, 4, 6)$, given eight stakeholders and $\boldsymbol{a} = (5, 6, \ldots, 12)$, $B = 100$. Note that the equity threshold solutions for $\Delta = (1, 3, 4, 6)$ are identical. A plot of the solutions for values of R (i.e., Δ) between 0 and 15 appears in Fig. 11.2.

13.3 Hierarchical Distribution

The problem of maximizing utility subject to a bound on utility range is not regionally decomposable. This is revealed by the example of the previous section, adapted to a two-level problem. Let region 1 consist of stakeholders 1 and 2, and region 2 consist of stakeholders 3 and 4. If we set $(B, B_1, B_2) = (0, 17\frac{1}{2}, 9\frac{1}{2})$ in the hierarchical problem, the collapsed problem is identical to the problem of the previous section with $B = 27$. The collapsed problem therefore has the solution $\bar{\boldsymbol{u}} = (\bar{\boldsymbol{u}}^1, \bar{\boldsymbol{u}}^2) = (2\frac{1}{2}, 2\frac{1}{2}, \frac{1}{2}, \frac{1}{2})$. Since $\boldsymbol{a}^1\bar{\boldsymbol{u}}^1 = 17\frac{1}{2} \geq B_1$ and $\boldsymbol{a}^2\bar{\boldsymbol{u}}^2 = 9\frac{1}{2} \geq B_2$, Proposition 3.1 implies that the hierarchical problem is collapsible and therefore has $\bar{\boldsymbol{u}}$ as an optimal solution. The region 1 problem has $k^* = 1$ and $R^* = 5\frac{5}{6}$, and since $R < R^*$, Proposition 13.1 implies that the region 1 problem has optimal solution $\hat{\boldsymbol{u}}^1 = (3\frac{9}{14}, 1\frac{9}{14})$. The region 2 problem has $k^* = 1$ and $R^* = \frac{1}{2}$, and since $R \geq R^*$, the problem has the solution $\hat{\boldsymbol{u}}^2 = (\frac{19}{14}, 0)$. Since $(\hat{\boldsymbol{u}}^1, \hat{\boldsymbol{u}}^2)$ differs from $\bar{\boldsymbol{u}}$ and is not optimal in the hierarchical problem, the latter is not regionally decomposable.

13.4 Incentives and Sharing

Marginal analysis of the bounded-inequality problem, subject to a linear budget constraint, is the same as for the equity threshold model, since the solution is the same. The details can be found in Chap. 11 with Δ replaced by R. Suffice it to say here that when $R \geq R^*$, only stakeholder k's utility is affected by efficiency improvements anywhere in the population. When $R < R^*$, an efficiency improvement by any stakeholder increases everyone's utility level by an equal absolute amount. Since raising the budget B can cause a transition from the former case to the latter, it can substantially boost the degree of benefit sharing as well as raise overall benefit levels.

Inequality Bounds: Gini Coefficient

14

14.1 Mathematical Formulation

Perhaps the best known and most frequently used inequality measure is the *Gini coefficient*. Partly for this reason, and partly due to the interesting and unexpected structure of the solution, we examine the problem of maximizing utility subject to a bound on the Gini coefficient. The coefficient is defined as

$$\frac{1}{2n^2\bar{u}}\sum_{ij}|u_i - u_j| = \frac{1}{n^2\bar{u}}\sum_{ij|i<j}|u_i - u_j| \tag{14.1}$$

Actually, we maximize utility subject to an unnormalized Gini statistic that is not divided by the mean utility $\bar{u}$, for the same reason as in the previous chapter. That is, maximizing the normalized statistic either yields a purely utilitarian solution or is infeasible. The unnormalized model nonetheless lends insight into the effect of maximizing welfare subject to a Gini-like inequality constraint. The model is

$$\max_{u,x}\left\{\sum_i u_i \;\middle|\; \frac{1}{n^2}\sum_{ij|i<j}|u_i - u_j| \leqslant G,\ (\boldsymbol{u},\boldsymbol{x}) \in S\right\} \tag{14.2}$$

where G is the desired bound on the unnormalized Gini coefficient.

The model (14.2) is easily linearized by introducing variables y_{ij} for $i < j$:

$$\max_{u,x,y}\left\{\sum_i u_i \;\middle|\; \frac{1}{n^2}\sum_{ij|i<j} y_{ij} \leqslant G;\quad \begin{array}{l} y_{ij} \geqslant u_i - u_j,\ y_{ij} \geqslant u_j - u_i,\ \text{all } i, j \text{ with } i < j \\ (\boldsymbol{u},\boldsymbol{x}) \in S \end{array}\right\}$$

Ö. Elçi et al., *The Structure of Fair Solutions*, Synthesis Lectures on Operations Research and Applications, https://doi.org/10.1007/978-3-031-82190-5_14

14.2 Socially Optimal Solution

Maximizing utility subject to a budget constraint and a Gini bound has a remarkably structured solution. It is similar to the solution obtained in the previous chapter subject to a range constraint (a much simpler constraint) in that there are one, two, or three utility levels. The lowest level is again zero if there are three levels. Yet the middle echelon of the range-bounded solution is of limited interest, since it can contain only one stakeholder. All three levels of a Gini-bounded solution can consist of multiple stakeholders. Furthermore, there is a single level in the Gini solution only when all the stakeholders have identical efficiency. For practical purposes, then, there are either two or three utility levels.

One might argue on this basis that the existence of two or three fairly discrete socioeconomic classes in many societies is not only a historical pattern, but somehow embedded in the mathematics of welfare maximization. This is perhaps a pessimistic outcome, in that it suggests that classed societies of this kind may be difficult to avoid. Yet one can adjust problem parameters so as to eliminate the lowest of the three classes and reduce the difference between the other two, while monitoring the effect on total utility. In particular, one can tighten the inequality bound and modify the individual conversion efficiencies. The latter can be accomplished by investing in education and other social benefits that tend to equalize the ability of individuals to generate utility with a given amount of resources.

We begin with two lemmas that simplify the argument to follow.

Lemma 14.1 *If $a_1 \leqslant \cdots \leqslant a_n$, there is an optimal solution $\boldsymbol{u}$ of (14.2) for which $u_1 \geqslant \cdots \geqslant u_n$.*

Proof Consider any optimal solution $\boldsymbol{u}$ of (14.2). If $u_i < u_j$ for some i, j with $i < j$, then let $\boldsymbol{u}'$ be a solution that is identical to $\boldsymbol{u}$ except that u_i and u_j swap values. There is no change in the objective function value, and the unnormalized Gini coefficient is the same, so that $\boldsymbol{u}'$ satisfies the inequality constraint. Also, $\boldsymbol{a}^\top \boldsymbol{u}' \leqslant \boldsymbol{a}^\top \boldsymbol{u} \leqslant B$ because $a_i \leqslant a_j$, which means that $\boldsymbol{u}'$ satisfies the budget constraint as well. The solution $\boldsymbol{u}'$ is therefore optimal. We now let $\boldsymbol{u} = \boldsymbol{u}'$ and repeat the argument until $u_1 \geqslant \cdots \geqslant u_n$. □

Lemma 14.2 *Given that $u_1 \geqslant \cdots \geqslant u_n$, the constraint*

$$\frac{1}{n^2} \sum_{ij|i<j} |u_i - u_j| \leqslant G \tag{14.3}$$

can be written as

$$\frac{1}{n^2} \sum_i (n - 2i + 1) u_i \leqslant G$$

Proof We first note that since $u_1 \geqslant \cdots \geqslant u_n$, we may replace $|u_i - u_j|$ in (14.3) with $u_i - u_j$. Now we observe that a given u_i occurs positively $n - i$ times and negatively $i - 1$ times in the summation. The coefficient of u_i is therefore $(n - i) - (i - 1) = n - 2i + 1$. □

Due to Lemmas 14.1 and 14.2, the problem (14.2) can be written as

$$\max \sum_i u_i \text{ subject to:} \tag{14.4}$$

$$\boldsymbol{a}^{\mathrm{T}}\boldsymbol{u} \leqslant B \tag{14.5}$$

$$\sum_i (n - 2i + 1)u_i \leqslant n^2 G \tag{14.6}$$

$$u_i \geqslant u_{i+1}, \; i = 1, \ldots, n-1 \tag{14.7}$$

$$\boldsymbol{u} \geqslant \boldsymbol{0} \tag{14.8}$$

Proposition 14.1 *There exists a utility distribution* $\boldsymbol{u}$ *that maximizes total utility, subject to a budget constraint and an upper bound* $G > 0$ *on the unnormalized Gini coefficient, in which the entire budget* B *is consumed and the utilities* u_i *take at most three different values. Furthermore, if there are exactly three utility levels, the smallest is zero. There is a single utility level only if* $a_1 = \cdots = a_n$, *in which case* $u_i = B/a(n)$ *for* $i = 1, \ldots, n$.

Proof To show that there is an optimal solution with at most three distinct utility levels, we show that an optimal basic solution must have at most three levels. Since at least one basic solution is optimal, the claim follows. We first rewrite the constraint set in (14.5)–(14.8) using equality constraints:

$$\boldsymbol{a}^T\boldsymbol{u} = B \tag{14.9}$$

$$\sum_i (n - 2i + 1)u_i + s_0 = n^2 G \tag{14.10}$$

$$-u_i + u_{i+1} + s_i = 0, \; i = 1, \ldots, n-1 \tag{14.11}$$

$$\boldsymbol{u}, \boldsymbol{s} \geqslant \boldsymbol{0} \tag{14.12}$$

where $\boldsymbol{s} = (s_0, \ldots, s_{n-1})$ are introduced as slack variables. We can write (14.9) without a slack variable, so that the entire budget is consumed, since any optimal solution satisfies (14.5) as an equation. This is because a solution $\boldsymbol{u}$ that results in slack in (14.5) can be improved by increasing each u_i by an equal amount without changing the left-hand side of (14.6). Recalling from Lemma 14.1 that we can assume $u_1 \geqslant \cdots \geqslant u_n$ in an optimal solution, we suppose that variables $u_1, \ldots, u_p$ are positive (and therefore basic) in a given basic solution, and that they occur on L different levels. Then L of the slack variables $s_1, \ldots, s_{n-1}$ are positive and therefore basic. This means that the $L + p$ columns of the coefficient matrix corresponding to these variables and $u_1, \ldots, u_p$ form part of a basis. However, only the first $2+p$ elements of these columns can be nonzero, which means that these columns can have rank at most $2+p$. Thus, they can form part of a basis only if $L+p \leqslant 2+p$, or $L \leqslant 2$, meaning that there can be at most two distinct utility levels among the positive u_is. This implies that there can be at most three utility levels (possibly including zero), and that the lowest level is zero if there are three.

To show that the solution is single tier only if $a_1 = \cdots = a_n$, we associate dual variable λ with constraint (14.5), γ with (14.6), and $\mu_1, \ldots, \mu_{n-1}$ with the constraints in (14.7). Then the dual of (14.4)–(14.8) is

$$\min_{\lambda,\gamma,\mu} \left\{ B\lambda + n^2 G\gamma \;\middle|\; \begin{array}{l} a_1\lambda + (n-1)\gamma - \mu_1 \geqslant 1 \\ a_i\lambda + (n-2i+1)\gamma + \mu_{i-1} - \mu_i \geqslant 1, \; i = 2, \ldots, n-1 \\ a_n\lambda + (1-n)\gamma + \mu_{n-1} \geqslant 1 \\ \lambda \geqslant 0, \; \gamma \geqslant 0, \; \mu_i \geqslant 0, \; i = 1, \ldots, n-1 \end{array} \right\}$$

If the solution is single tier, so that each $u_i > 0$, the first n dual constraints are satisfied as equations due to complementary slackness. Summing these equations, we obtain $\lambda = n/a(n)$. Furthermore, the left-hand side of constraint (14.6) vanishes, in which case the constraint is slack, and we must have $\gamma = 0$. Using backsubstitution, the dual constraints and $\gamma = 0$ imply that

$$\mu_i = a(i)\lambda - i = n\frac{a(i)}{a(n)} - i, \quad i = 1, \ldots, n-1 \tag{14.13}$$

But a dual feasible solution must have $\mu_i \geqslant 0$ for all i, which due to (14.13) and $a_1 \leqslant \cdots \leqslant a_n$ is possible only if $a_1 = \cdots = a_n$. Thus a single-tier solution is optimal only if $a_1 = \cdots = a_n$. Constraint (14.5) implies that the solution must be $u_i = B/a(n)$ for all i. □

A complete description of the optimal solution can, in principle, be stated, but it yields little further insight into the problem because it depends on a number of complex conditions. In any event, an optimal solution is easily obtained in any given instance by solving the linear programming problem (14.4)–(14.8).

As an example, suppose we wish to maximize total utility $u_1 + \cdots + u_5$ subject to $3u_1 + 4u_2 + 6u_3 + 7u_4 + 10u_5 \leqslant B$ with $B = 30$. If the upper bound G on the unnormalized Gini coefficient satisfies $G \geqslant 1.6$, we obtain the purely utilitarian solution $\boldsymbol{u} = (10, 0, 0, 0)$. If $0.24 < G < 1.6$, the solution has three tiers. For example, with $G = 1.2$ we obtain $\boldsymbol{u} = (6, 3, 0, 0)$, and with $G = 0.6$ we obtain $\boldsymbol{u} = (\frac{255}{92}, \frac{255}{92}, \frac{75}{92}, \frac{75}{92}, 0) \approx (2.772, 2.772, 0.815, 0.815, 0)$. Note that the middle tier can consist of more than one individual, unlike the solution of a bounded range problem. Setting $G = 0.24$ yields another two-tier solution with the bottom tier equal to zero: $\boldsymbol{u} = (\frac{3}{2}, \frac{3}{2}, \frac{3}{2}, \frac{3}{2}, 0)$. If $0 < G < 0.24$, we obtain all-positive two-tier solutions. For example, with $G = 0.12$ we obtain $\boldsymbol{u} = (\frac{5}{4}, \frac{5}{4}, \frac{5}{4}, \frac{5}{4}, \frac{1}{2})$.

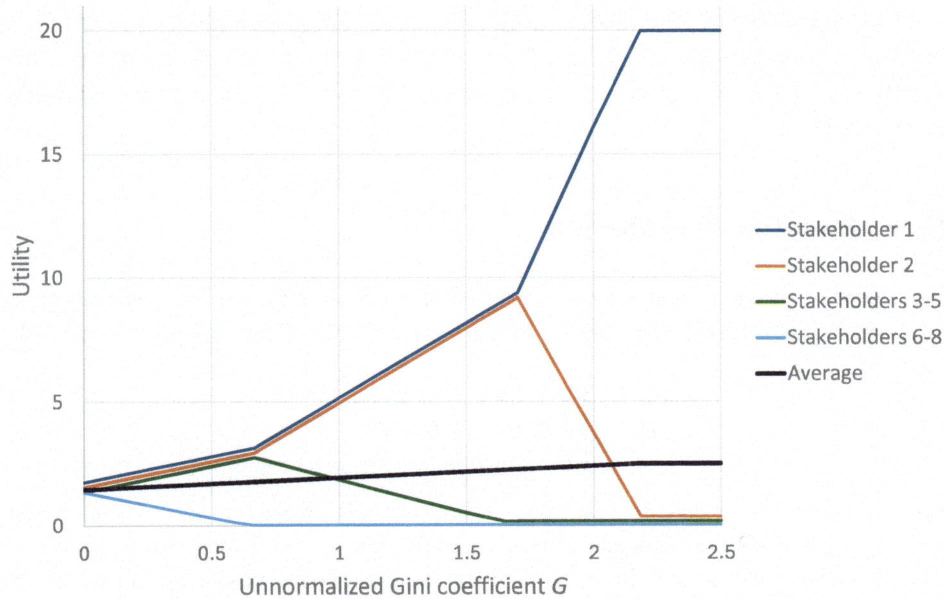

Fig. 14.1 Utility distribution over eight stakeholders that maximizes total utility subject to an unnormalized Gini bound. Lines that coincide are slightly offset to clarify the pattern

Table 2.2 indicates solutions corresponding to $G = 0.5, 1, 1.5, 2$ when there are eight stakeholders with $\boldsymbol{a} = (5, 6, \ldots, 12)$ and $B = 100$. Figure 14.1 is a plot of solutions for the same problem instance corresponding to values of G between 0 and 2.5. Both the table and the plot illustrate that the middle tier can contain multiple stakeholders for certain values of G. In the plots, for example, stakeholders 3, 4, and 5 comprise the middle tier when G is between approximately 0.67 and 1.7.

14.3 Hierarchical Distribution

The problem of maximizing utility subject to a bound on the unnormalized Gini coefficient is not regionally decomposable. This is shown by the example of the previous section, adapted to a two-level problem. Let region 1 consist of stakeholders 1 and 2, and region 2 consist of stakeholders 3, 4 and 5, and suppose $G = 0.12$. If we set $(B, B_1, B_2) = (0, 8\frac{3}{4}, 21\frac{1}{4})$ in the hierarchical problem, the collapsed problem is identical to the problem of the previous section. The collapsed problem therefore has the solution $\bar{\boldsymbol{u}} = (\bar{\boldsymbol{u}}^1, \bar{\boldsymbol{u}}^2) = (\frac{5}{4}, \frac{5}{4}, \frac{5}{4}, \frac{5}{4}, \frac{1}{2})$. Since we have $\boldsymbol{a}^1\bar{\boldsymbol{u}}^1 = 16\frac{1}{4} \geqslant B_1$ and $\boldsymbol{a}^2\bar{\boldsymbol{u}}^2 = 13\frac{3}{4} \geqslant B_2$, Proposition 3.1 implies that the hierarchical problem is collapsible and therefore has $\bar{\boldsymbol{u}}$ as an optimal solution. The region 1 problem has the optimal solution $\hat{\boldsymbol{u}}^1 = (\frac{1067}{700}, \frac{731}{700}) \approx (1.524, 1.044)$, and the region 2 problem has the solution $\hat{\boldsymbol{u}}^2 = (\frac{533}{460}, \frac{533}{460}, \frac{1423}{2300}) \approx (1.159, 1.159, 0.619)$. Since $(\hat{\boldsymbol{u}}^1, \hat{\boldsymbol{u}}^2)$ differs from $\bar{\boldsymbol{u}}$ and is not optimal in the hierarchical problem, the latter is not regionally decomposable.

14.4 Incentives and Sharing

The incentive structure of the problem can be analyzed in terms of the current optimal solution $\boldsymbol{u}^*$. Since the solution consists of two or three tiers, we can suppose that it has the form:

$$u_i^* = \begin{cases} v^*, & \text{for } i = 1, \ldots, k \\ w^*, & \text{for } i = k+1, \ldots, \ell \\ 0, & \text{for } i = \ell+1, \ldots, n \end{cases}$$

where $1 \leqslant k \leqslant \ell \leqslant n$ and $v^* > w^* > 0$. There are only two tiers when $k < \ell = n$ or $k = \ell < n$. The analysis depends in rather complicated ways on the relative values of B, G, k, ℓ and n, but to simplify matters we assume the common situation in which the fairness constraint (14.6) is binding, $k < \ell$, and $k + \ell > n$. The latter inequality implies that most of the utilities are positive.

The perturbation analysis assumes that the current optimal basis remains optimal. Since there are only two nonzero utility values v^* and w^*, the current basic solution is determined (given k and ℓ) by the two equations (14.5) and (14.6). These equations can be written as

$$\begin{aligned} &a(n)v^* + a(k{+}1{:}\ell)w^* = B \\ &k(n-k)v^* - (\ell-k)(k+\ell-n)w^* = n^2G \end{aligned}$$

Their solution is

$$v^* = \frac{B + \dfrac{n^2G}{(\ell-k)(k+\ell-n)}a(k{+}1{:}\ell)}{a(k) + \dfrac{k(n-k)}{(\ell-k)(k+\ell-n)}a(k{+}1{:}\ell)}$$

$$w^* = \frac{B - \dfrac{n^2G}{k(n-k)}a(k)}{\dfrac{(\ell-k)(k+\ell-n)}{k(n-k)}a(k) + a(k{+}1{:}\ell)}$$

The cross-substitution rate from a stakeholder i to stakeholders in the upper tier is therefore

$$-\frac{\partial v^*}{\partial a_i}\cdot\frac{a_i}{v^*}=\begin{cases}\dfrac{a_i}{a(k)+\dfrac{k(n-k)}{(\ell-k)(k+\ell-n)}a(k{+}1{:}\ell)}, & \text{for } i=1,\dots,k\\[2ex] \dfrac{\left(1-\dfrac{1}{v^*}\cdot\dfrac{n^2G}{k(n-k)}\right)a_i}{\dfrac{(\ell-k)(k+\ell-n)}{k(n-k)}a(k)+a(k{+}1{:}\ell)}, & \text{for } i=k+1.\dots,\ell\\[2ex] 0, & \text{for } i=\ell+1,\dots,n\end{cases}$$

This implies that a stakeholder in the upper tier benefits from efficiency improvements of any stakeholder i in the upper tier, in an amount roughly proportional to i's utility cost a_i. There is no benefit from improvements in the lowest tier, where the utility allotments are zero. The benefit from efficiency improvements in the middle tier is positive if the numerator of the fraction defining the cross-subsidy rate is positive, which is to say $k(n-k)v^* > n^2D$. Using the solution value for v^* above, this condition is equivalent to

$$B > \frac{n^2G}{k(n-k)}a(k) \tag{14.14}$$

Thus, there is sharing with the upper tier when the emphasis on fairness is sufficiency great (i.e., G is sufficiently small), and competition otherwise, as one might expect.

The cross-substitution rate from a stakeholder i to stakeholders in the middle tier is

$$-\frac{\partial w^*}{\partial a_i}\cdot\frac{a_i}{w^*}=\begin{cases}\dfrac{\left(1+\dfrac{1}{w^*}\cdot\dfrac{n^2G}{(\ell-k)(k+\ell-n)}\right)a_i}{a(k)+\dfrac{k(n-k)}{(\ell-k)(k+\ell-n)}a(k{+}1{:}\ell)}, & \text{for } i=1,\dots,k\\[2ex] \dfrac{a_i}{\dfrac{(\ell-k)(k+\ell-n)}{k(n-k)}a(k)+a(k{+}1{:}\ell)}, & \text{for } i=k+1,\dots,\ell\\[2ex] 0, & \text{for } i=\ell+1,\dots,n\end{cases}$$

This indicates that the middle tier always receives a positive benefit from an efficiency improvement by a stakeholder i in either of the two positive tiers, with a magnitude again roughly proportional to a_i.

Finally, we note that all stakeholders who receive positive utility are incentivized to improve their own efficiency, and that incentive is greater for less efficient stakeholders.

To illustrate these results, let $G = 0.6$ in the above example, which yields the three-tiered utility distribution $\boldsymbol{u} \approx (2.772, 2.772, 0.815, 0.815, 0)$. The incentive rate for stakeholder 1 is $\frac{3}{46} \approx 0.065$, meaning that the stakeholder reaps approximately a 0.065% increase in allocated utility for each 1% improvement in its efficiency. The incentive rate for the less efficient stakeholder 3 is $\frac{9}{23} \approx 0.391$. The incentive rates are rather small, due to the fairness setting $G = 0.6$, but positive nonetheless. The cross-subsidy rates from the middle to the top tier are positive, because (14.14) is satisfied. For example, the cross-subsidy rate for stakeholder 3 to either member of the top tier is $\frac{15}{391} \approx 0.038$. Setting $G > \frac{36}{35}$ results in a negative subsidy. As noted earlier, cross-subsidy rates from the top to the middle tier are always positive. For example, the rate from stakeholder 1 to the middle tier is $\frac{3519}{5290} \approx 0.665$, substantially larger than subsidy rates in the opposite direction.

Conclusions 15

15.1 Balancing Equity and Efficiency

This final chapter attempts to construct a narrative from the foregoing analysis that can help one select the right fairness criterion. The first lesson to be learned from this analysis is that a purely efficiency-oriented criterion, or even a purely equity-oriented criterion, is likely to be unsatisfactory for most applications where fairness is a concern. Some means of balancing the two is normally required. The aim of this book has been to examine the kind of balance achieved by various social welfare functions, by means of a structural analysis of the solutions they deliver. Tables 15.1, 15.2, 15.3 and 15.4 summarize the results of this inquiry.

A *utilitarian* objective, which maximizes total utility, focuses entirely on efficiency. When utility is maximized subject to a simple budget constraint, a single stakeholder receives all available resources and others receive nothing, an outcome that is only marginally ameliorated by placing bounds on resource allotments. While this extreme result is not evident in most practical optimization models, due to the complexity of the constraint set, this complexity only serves to conceal the potential for inequity in a purely utilitarian criterion.

An objective based primarily on equity can yield equally extreme solutions. Perhaps the most famous fairness criterion, the *maximin* objective that derives from the Rawlsian difference principle, forces all parties to accept the same level of utility, except, again, where this is blocked by other constraints—constraints that may reflect only the situation and no coherent understanding of what is just. For example, if one person is difficult to accommodate, due to unfortunate circumstances such as an incurable disease, it is necessary to lavish resources on that person to the point that all others are reduced to the same minimal utility gain. Even if we prevent this outcome by placing an upper bound on the disadvantaged party's utility, the maximin objective allows us to allot others that same low level of utility, when they could receive much more. A *leximax* objective largely removes this second anomaly, but the extreme solutions remain.

Ö. Elçi et al., *The Structure of Fair Solutions*, Synthesis Lectures on Operations Research and Applications, https://doi.org/10.1007/978-3-031-82190-5_15

15.2 Socially Optimal Distributions

We now proceed to examine the solutions obtained by social welfare functions that combine equity and efficiency in a more balanced fashion. The structural results are summarized in Tables 15.1 and 15.2. Results related to hierarchical distributions and marginal analysis are summarized in subsequent sections.

Table 15.1 Structural characteristics of optimal distributions for various social welfare functions, part 1

Social welfare criterion	Solution subject to simple budget constraint	Other structural properties
Utilitarian	Most efficient stakeholder receives all resources, others receive nothing. No concern for equity	With upper bounds on utilities, more efficient stakeholders collectively receive everything, others nothing
Maximin	Available resources distributed so that all stakeholders receive equal utility. All stakeholders reduced to the level of disadvantage of the worst off. Solution can be highly disutilitarian	With upper bounds on utility bounds, some resources can be unused. When one stakeholder has a tight bound, the equity situation of other disadvantaged stakeholders is not considered
Leximax	Same solution as maximin, even if it is highly disutilitarian	No resources are left unused. Regardless of constraints, the equity situation of all disadvantaged stakeholders is considered
Beta fairness	Same as maximin solution, thus defeating the purpose of beta fairness, unless some stakeholders are much less efficient than others, in which case they receive no utility	Solution can be other than maximin if β is close to 1. So beta fairness may be better suited to avoid a small overprivileged group at the top rather than a small underprivileged group at the bottom
Alpha fairness	Closed-form solution. All stakeholders receive some utility when $\alpha > 0$; more efficient stakeholders receive more, but with greater equity for larger α. Maximin solution when $\alpha = \infty$	Solutions with infinite inequality can have same social welfare values as egalitarian solutions. Proportional fairness ($\alpha = 1$) subject to a simple budget constraint distributes utilities in proportion to stakeholder efficiency
Kalai–Smorodinsky bargaining	Each stakeholder receives equal fraction of its ideal utility (i.e., utility it would experience if others received no resources). Same solution as alpha fairness for $\alpha = 1$ (proportional fairness)	With upper bounds on utilities, each stakeholder receives an equal fraction of its upper bound. If ideal utilities are unchanged, enlarging the feasible set never reduces any stakeholder's utility share

Table 15.2 Structural characteristics of optimal distributions for various social welfare functions, part 2

Social welfare criterion	Solution subject to simple budget constraint	Other structural properties
Utility threshold with maximin	Only 2 possible solutions: purely utilitarian for Δ less than a calculable value, purely maximin for larger Δ	With utility bounds, solution is generally neither utilitarian nor maximin. At most one utility is strictly between its upper bound and the smallest utility
Equity threshold with maximin	For larger Δ (less emphasis on fairness), more efficient stakeholders receive utility Δ, others (with 1 possible exception) receive none	For small Δ, all stakeholders receive utility, and more efficient stakeholders receive equal utility that is Δ more than that received by less efficient
Utility threshold with leximax	More efficient stakeholders receive equal utility (their number is larger for larger Δ), less efficient receive none	More sensitive to equity of all disadvantaged stakeholders (not just very worst off) than is the utility threshold criterion with maximin
Max utility with range bound	Same solution as the equity threshold problem with maximin when Δ is set equal to the upper bound on the utility range	Less general than the equity threshold problem with maximin, which can deliver solutions with utility range greater than Δ when solved subject to constraints other than a simple budget constraint
Max utility with Gini bound	Utility values in the solution must occur on 2 or 3 levels, with all utilities equal in each level. If there are 3 levels, all utilities in one level are zero	May partially explain historical occurrence of societies with 2 or 3 major socioeconomic classes

A natural way to avoid the extreme solutions of maximin criterion is to consider the welfare of a disadvantaged *class* of stakeholders rather than only the very worst-off stakeholder. *Beta fairness* accomplishes this by maximizing the average of the βn smallest utilities, where n is the number of stakeholders. One would normally set β to a fairly small value (perhaps between 0.1 and 0.5) to focus attention on the disadvantaged. However, this typically yields the same solution as a maximin criterion (subject to a linear budget constraint), thus defeating the purpose of beta fairness. Beta fairness may therefore be more useful for avoiding an overprivileged class (by settling β closer to 1) than for avoiding an underprivileged class.

Alpha fairness, perhaps the best known composite criterion, allows a parameterized trade-off between fairness and efficiency. The solution transitions smoothly from pure utilitarian when $\alpha = 0$ to pure maximin as $\alpha \to \infty$, with intermediate values of α yielding solutions that balance equity and efficiency. Setting $\alpha = 1$ results in *proportional fairness* (also known as the *Nash bargaining solution*), which in the presence of a simple budget constraint allocates utility to stakeholders in proportion to their conversion efficiency. Alpha fairness has seen a number of applications in engineering, especially telecommunications. However, there is no clear theoretical guidance for selecting a suitable α in a given context. With experience, a community can settle on a benchmark value of α in a particular problem domain, much as is done for confidence intervals or p-values in statistics. Future applications can then be meaningfully compared with past experience with respect to the degree of fairness achieved. Oddly, when the resource constraints are more complicated than a simple budget constraint, alpha fairness for $\alpha \geqslant 1$ can regard an egalitarian distribution as no better than one in which there is arbitrarily extreme inequality. Alpha fairness, and in particular the Nash bargaining solution, also has a curious nonmonotonicity property. It can reduce the utility allocation to some stakeholders when more resources become available, an outcome these stakeholders could regard as unacceptable.

Kalai–Smorodinsky bargaining was introduced, in part, to modify Nash bargaining so as to avoid the nonmonotonicity property just mentioned. It takes into account the maximum utility each stakeholder could experience if allocated all available resources. It then maximizes the fraction of this maximum that each stakeholder receives, subject to the condition that the fraction is the same for all stakeholders. It therefore assesses fairness while taking into account the cost of delivering utility to each stakeholder. This may be a desirable feature in bargaining, as when labor and management negotiate a contract, since either side may regard an agreement as unfair if the other side receives a larger share of its ideal outcome. It may also be suitable when some clients are more difficult to service than others, as when less frequent bus service is provided to sparsely populated areas due to the greater cost of reaching them. On the other hand, it cannot compensate for bad luck, because it awards privileged or more fortunate stakeholders the same fraction of their potential utility as those who have far less potential due to some physical or mental impairment beyond their control. In the presence of a simple budget constraint, the K–S solution is identical to the Nash bargaining solution (alpha fairness with $\alpha = 1$). Unlike alpha fairness, it allows for no adjustments in the equity/efficiency trade-off. In addition, increasing available resources cannot reduce any stakeholder's utility allotment, provided there is no change in the maximum utility that any stakeholder can receive.

Threshold functions provide an alternate means for combining equity and efficiency, where the balance is governed by a parameter Δ that may be more easily interpreted in practice than the α in alpha fairness. A *utility threshold* function uses a maximin criterion until the utility cost crosses a threshold, at which point it begins to switch to a utilitarian criterion. It can be appropriate to applications in which equity is the primary concern, but not to the point of sacrificing excessive utility. The equity/utility trade-off is governed by a parameter Δ that delivers a purely utilitarian solution when $\Delta = 0$ and a maximin solution as $\Delta \to \infty$. Stakeholders whose utility level is within Δ of the lowest utility are regarded as disadvantaged and are given additional weight when assessing social welfare. As it turns out, the utility threshold solution is purely utilitarian or purely maximin when obtained subject to a simple budget constraint, although one can state in closed form which values of Δ produce one or the other. The addition of utility bounds results in solutions with a more interesting structure that can provide managerial insights. In particular, at most one stakeholder has a utility level that lies strictly between its upper bound and the smallest utility.

An *equity threshold* function is utilitarian until inequities become too great, at which point it begins to switch to a maximin criterion. It may therefore be suitable for applications in which efficiency is the primary concern, but without too much sacrifice of equity. The trade-off is again regulated by a parameter Δ, but in the opposite direction; smaller values of Δ result in a greater emphasis on fairness. The interpretation of Δ is also quite different. Stakeholders whose utility is within Δ of the lowest are again regarded as disadvantaged, but increasing the utility of more fortunate stakeholders adds nothing to social welfare unless the lowest utility is increased by an equal amount. The socially optimal solution subject to a simple budget constraint is somewhat more interesting than that for a utility threshold SWF. When Δ is set near the utilitarian end of the scale, it awards a utility level of Δ to several of the more efficient stakeholders (their number depending on Δ) and nothing to the remainder. When Δ is set to emphasize equity, it awards everyone positive utility, but the more efficient stakeholders receive Δ more than the others.

These two threshold functions inherit a shortcoming of the maximin criterion, if only in attenuated form. This is the tendency to give insufficient attention to disadvantaged parties other than the very worst off. This may not be a serious issue when maximizing the lowest utility tends to push up other smaller utilities as well. However, when there is a tight upper bound on the smallest utility that prevents it from being raised above a low level, then the equity situation of other less advantaged stakeholders is ignored. This situation can be remedied by a threshold function that combines a utilitarian objective with a *leximax* rather than a maximin criterion.

A *utility threshold criterion with leximax*, for example, can be implemented by solving a certain sequence of optimization problems. In contrast to a utility threshold function with maximin, it can recommend solutions (subject to a simple budget constraint) that lie between the utilitarian and leximax extremes. Specifically, a socially optimal solution allocates equal utility to some of the more efficient stakeholders and none to the remaining stakeholders. Larger values of Δ (which correspond to greater equity) extend positive benefits to more stakeholders, with a correlative reduction in the size of the benefit. The solutions are more nuanced in the presence of utility bounds, and they incorporate fairness to all disadvantaged stakeholders rather than only the very worst off. An *equity threshold function with leximax* has apparently not been studied and poses an interesting research problem.

An alternative approach to balancing equity and efficiency is to maximize total utility subject to a bound on some kind of equity-sensitive social welfare function. A difficulty with this type of model is that the SWF provides no guidance as to how it should be balanced with the utilitarian objective. This contrasts with the models so far discussed, which minimize a SWF directly and therefore compel one to embed the equity/efficiency balancing scheme transparently in the SWF where it can be defended or questioned. We nonetheless consider two models that maximize utility subject to a bound on inequality, because they yield very interesting results. One measures inequality by the range of utilities, and the other by an unnormalized Gini coefficient.

A model that maximizes utility subject to an upper bound on the *utility range* turns out to be closely related to the equity threshold criterion discussed earlier. In fact, the two models deliver exactly the same solution when solved subject to a simple budget constraint, provided the parameter Δ in the threshold model is set equal to the maximum desired utility range. The threshold model is more general, however, in that it can deliver solutions in which the utility range is greater than Δ in the presence of other types of constraints. This suggests that the threshold model is preferable due to its greater flexibility.

When inequality is measured by the widely used *Gini coefficient*,[1] the resulting socially optimal solution exhibits a remarkable structure. It partitions the stakeholders into two or three groups, within each of which everyone receives the same utility allotment. When there are three groups, stakeholders in one group receive zero utility. In other words, a society that maximizes utility subject to a Gini-based inequality bound and a linear budget constraint necessarily consists of two or three distinct and homogeneous socioeconomic classes. Fortunately, differences between these classes can be reduced by imposing a stricter inequality bound, or by providing stakeholders with educational and other benefits that tend to equalize their conversion efficiencies (i.e., their ability to convert resources to utility).

15.3 Hierarchical Distributions

The foregoing narrative can be enriched by observing the behavior of the various social welfare criteria on *hierarchical networks* in which resources can be passed from a national to a regional level. Each region combines its own resources with those received from above and distributes them to subregions (which can be interpreted as schools, hospitals, or other institutions). All of the extremes discussed above persist in this context. However, the simpler models are more likely to be regionally decomposable, due to a technical property (monotone separability) of the corresponding social welfare functions. Regional decomposability means that if each region distributes its resources using the same social welfare criterion as used at the national level, it will obtain an allocation to subregions that is consistent with that prescribed at the national level. The allocation computed by the national authority assumes, in fact, that the allocation within regions will follow this pattern. When there is no regional decomposability, the national solution is valid only if the regions follow the national prescription rather than re-solving the allocation problem for their region alone.

[1] The coefficient is not normalized by the mean utility, as is the usual practice, because the resulting model is useless for achieving fairness, as explained in Chapter 14.

Table 15.3 Hierarchical and marginal analysis of various social welfare functions, when maximized subject to a linear budget constraint, part 1

Social welfare criterion	Regionally decomposable?	Incentives and sharing
Utilitarian	Yes	Only most efficient stakeholder incentivized to improve efficiency, no sharing
Maximin	Yes	Less efficient stakeholders have greater incentive to improve efficiency. Benefits of improvement are shared equally* with other stakeholders
Leximax	Yes	Same as maximin
Beta fairness	No	Less efficient stakeholders who are currently allocated positive utility have greater incentive to improve efficiency. The benefits of improvement are shared equally* with other stakeholders who are allocated positive utility
Alpha fairness	Yes	Less efficient stakeholders have greater incentive to improve efficiency. Competition is incentivized for $\alpha < 1$ (efficiency improvements take utility from others). Cooperation is incentivized for $\alpha > 1$ (benefits of efficiency improvements are shared with others). Proportional fairness ($\alpha = 1$) results in no cross subsidies
Kalai–Smorodinsky bargaining	Yes, if collapsible, for which there is a simple test	Same as proportional fairness (alpha fairness with $\alpha = 1$)

* Benefits are shared equally in the sense that each stakeholder's share is increased by the same percentage

The specific results are summarized in Tables 15.3 and 15.4. The purely utilitarian, maximin, and leximax criteria are regionally decomposable, as is alpha fairness. The Kalai–Smorodinsky model is regionally decomposable when it is collapsible, meaning that the multilevel problem can be solved as a single-level problem. A sufficient condition for collapsibility can be checked by applying a simple test. The utility and equity threshold models can be regionally nondecomposable even when they are collapsible. Models that maximize utility subject to a utility range or Gini bound likewise lack decomposability.

One might infer from these results that the more complex models are progressively less prone to be decomposable. Lack of regional decomposability might be regarded as an

Table 15.4 Hierarchical and marginal analysis of various social welfare functions, when maximized subject to a linear budget constraint, part 2

Social welfare criterion	Regionally decomposable?	Incentives and sharing
Utility threshold with maximin	No	Same as utilitarian or maximin, depending on Δ
Equity threshold with maximin	No	For larger Δ, only 1 stakeholder is incentivized to improve, and this stakeholder benefits from improvements by more efficient stakeholders. For smaller Δ, all are incentivized to improve, benefits shared with all
Utility threshold with leximax	No	Only stakeholders allocated positive utility have incentive to improve (less efficient have greater incentive), with benefits shared equally* with other stakeholders allocated positive utility
Max utility with range bound	No	Same as equity threshold, with Δ set equal to the utility range bound *R*
Max utility with Gini bound	No	Less efficient stakeholders who receive positive utility have greater incentive to improve efficiency. If there are 3 utility tiers, top tier benefits from efficiency improvements in that tier and possibly in the middle tier (depending on a computable problem data statistic). Middle tier benefits from efficiency improvements in itself and the top tier. Lowest tier benefits from no one's efficiency improvements

* Benefits are shared equally in the sense that each stakeholder's share is increased by the same percentage

inconvenience, but it can also signal greater adequacy and subtlety as an equity measure. Perhaps local decisions should reflect a larger perspective if they are to be truly fair.

15.4 Incentives and Sharing

When selecting a social welfare function, it may be important to consider the *incentives* that various SWFs provide for stakeholders to invest in efficiency improvements. If there is an interest in promoting cooperation or competition, one can also take note of the extent to which the fruits of these improvements are *shared* with other stakeholders.

These effects can be determined by marginal analysis of the socially optimal solution. Incentives can be measured by a stakeholder's percentage gain in utility, at the margin, for a given percentage improvement in that stakeholder's efficiency improvement. The degree of sharing is indicated by cross subsidies, which can be measured by a stakeholder's percentage gain (or loss) in utility for a given percentage improvement in another stakeholder's efficiency. A positive cross subsidy incentivizes cooperation, because a stakeholder benefits from helping others to be more efficient. A negative cross subsidy incentivizes competition, because a stakeholder who improves efficiency takes resources from others.

The results of marginal analysis for the various SWFs, subject to a linear budget constraint, are summarized in Tables 15.3 and 15.4. A purely *utilitarian* SWF provides an incentive only to the single stakeholder that is allocated positive utility in the socially optimal solution. By contrast, a purely *maximin* SWF provides an incentive to every stakeholder, and less efficient stakeholders have a proportionately stronger incentive to improve. The benefits of one stakeholder's improvement are equally shared with all stakeholders, in the sense that each stakeholder's utility is increased by the same percentage. Marginal analysis for a *leximax* criterion is the same as for maximin. *Beta fairness* is very similar to maximin in its behavior at the margin, except that only stakeholders who currently receive positive utility are incentivized to improve efficiency, and only they benefit from the efficiency improvements of others.

Alpha fairness provides an incentive to improve efficiency that is greater for less efficient stakeholders. However, the incentives are competitive in nature for $\alpha < 1$ and cooperative for $\alpha > 1$. For $\alpha < 1$, a stakeholder who improves efficiency receives additional utility at the expense of the other stakeholders. For $\alpha > 1$, part of the benefit of improved efficiency is shared with all other stakeholders. For $\alpha = 1$ (proportional fairness), there are no cross subsidies in either direction. These facts can help inform the choice of α in a particular situation. The *Kalai–Smorodinsky* bargaining solution has the same incentive and sharing characteristics as proportional fairness.

Incentive characteristics of a *utility threshold* SWF with maximin fairness are the same as those of the utilitarian or maximin SWF, depending on the value of the trade-off parameter Δ. The marginal behavior of an *equity threshold* SWF with maximin fairness depends on its parameter Δ in a different way. For smaller (more equitable) values of Δ, it provides improvement incentives to all stakeholders, incentives that are greater for the less efficient, and the benefits of efficiency improvements are shared equally with all stakeholders. For larger (more utilitarian) values of Δ, only one stakeholder (one with an intermediate efficiency level) benefits from anyone's efficiency improvements. A *utility threshold* SWF with *leximax* fairness favors cooperation, but only among stakeholders who receive positive utility. The less efficient among these are more strongly incentivized to improve efficiency.

A model that maximizes utility subject to a bound on the *utility range* has the same marginal characteristics as the equity threshold model, provided the parameter Δ is set equal to the desired maximum utility range. When inequality is measured by an unnormalized *Gini coefficient* rather than the utility range, the marginal analysis is quite interesting. Less efficient stakeholders who currently receive positive utility have a greater incentive to improve efficiency. When there are three utility levels in the solution, cross subsidies can flow from the top tier to the middle tier and from one stakeholder to another in either of these tiers. Whether cross subsidies can flow from the middle tier to the top tier depends on problem parameters according to a simple formula. As one might expect, subsidies in this direction are more likely when there is a tighter constraint on inequality. Stakeholders in the bottom tier, who currently receive no utility, receive no cross subsidies from more privileged stakeholders.

15.5 A Final Observation

One can view the project undertaken in this book from a broader perspective. An *ex post* characterization of social welfare functions may have the potential to initiate a mutually beneficial dialog between optimization and ethics. There is precedent for this, because interaction between optimization and other areas has borne fruit in the past. A notable example is the intellectual cross fertilization between linear programming and duality theory on the one hand and economics and game theory on the other. Structural analysis of equitable distributions can help refine concepts of distributive justice in ethics, while ethical analysis can assist the mathematical formulation of fairness in optimization models.

Trade-Off Plots for Two Stakeholders

In this appendix, we explore graphically the influence of stakeholder conversion efficiencies on socially optimal solutions when there are two stakeholders. The optimization problem is to maximize a given social welfare function subject to the budget constraint $a_1u_1 + a_2u_2 \leq B$ and $u_i \geq 0$, where coefficient a_i corresponds to a conversion efficiency $1/a_i$. To simplify the presentation, we fix $a_1 = 1$ and allow a_2 to vary. We also set the budget $B = 1$.

The results appear in Figs. A.1, A.2, A.3, A.4, A.5, A.6, and A.7. The x-axis of each figure is stakeholder 2's coefficient a_2. For each social welfare function, we present two graphs: one showing the resulting utility allocation and the other showing the resulting resource allocation. These figures provide a minimal representation of what the resource and utility allocations would look like if a society consists of two equally sized groups of people with conversion efficiencies of 1 and $1/a_2$, respectively.

Utilitarian SWF (Fig. A.1): This graph shows that the utilitarian approach maximizes the total utility across all stakeholders without concern for equity. As Stakeholder 2's coefficient (a_2) increases (becoming less efficient in resource conversion), the resource allocation heavily favors the more efficient stakeholder (Stakeholder 2 initially and Stakeholder 1 later). This illustrates the key characteristic of utilitarianism. This can lead to highly unequal distributions, as the less efficient stakeholder receives minimal resources.

Maximin SWF (Fig. A.2): In contrast to utilitarianism, the maximin criterion focuses on improving the utility of the worst-off stakeholder. Here, we see a more balanced distribution between Stakeholder 1 and Stakeholder 2, with resources allocated in a way that guarantees a minimal level of utility for both stakeholders. As Stakeholder 2 becomes less efficient (a_2 increases), they begin receiving a greater share of the total resources. This reflects the maximin principle's concern for equity and its attempt to avoid extreme inequality.

Alpha Fairness SWF (Fig. A.3): With alpha set at 0.5, this SWF also balances efficiency and equity but places a stronger emphasis on fairness compared to beta fairness SWF with

Ö. Elçi et al., *The Structure of Fair Solutions*, Synthesis Lectures on Operations Research and Applications, https://doi.org/10.1007/978-3-031-82190-5

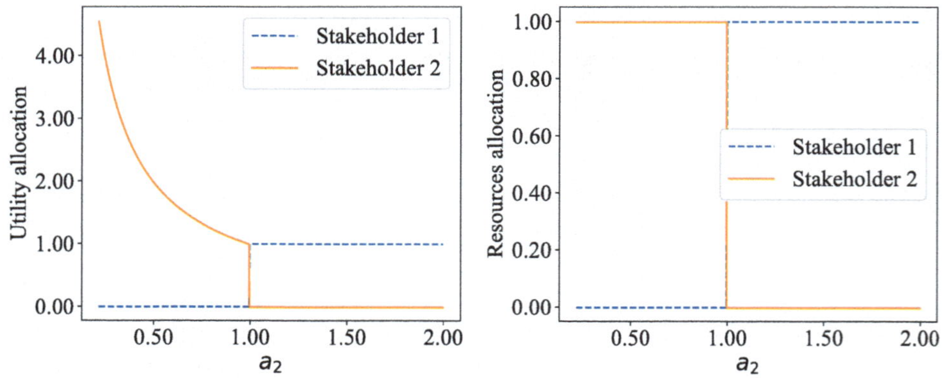

Fig. A.1 Utilitarian SWF. Total budget = 1. Stakeholder 1 coefficient $a_1 = 1$

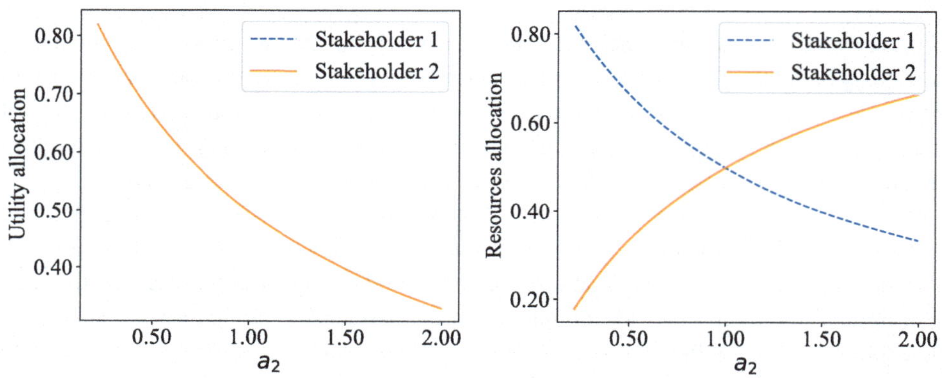

Fig. A.2 Maximin SWF. Total budget = 1. Stakeholder 1 coefficient $a_1 = 1$

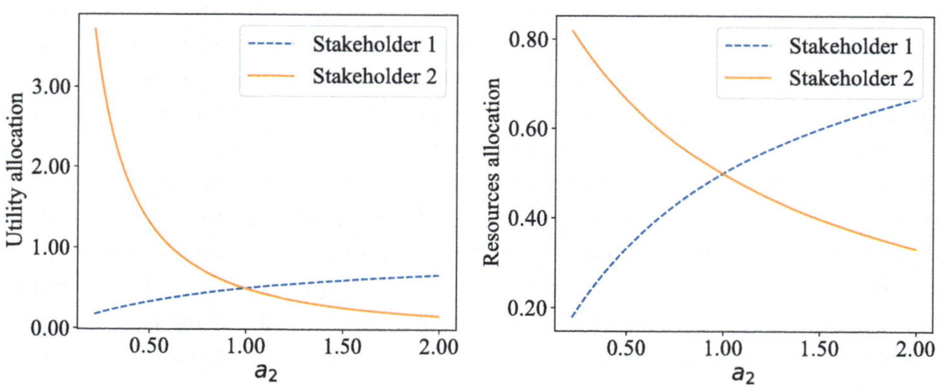

Fig. A.3 Alpha fairness SWF. Total budget = 1. Stakeholder 1 coefficient $a_1 = 1$. alpha = 0.5

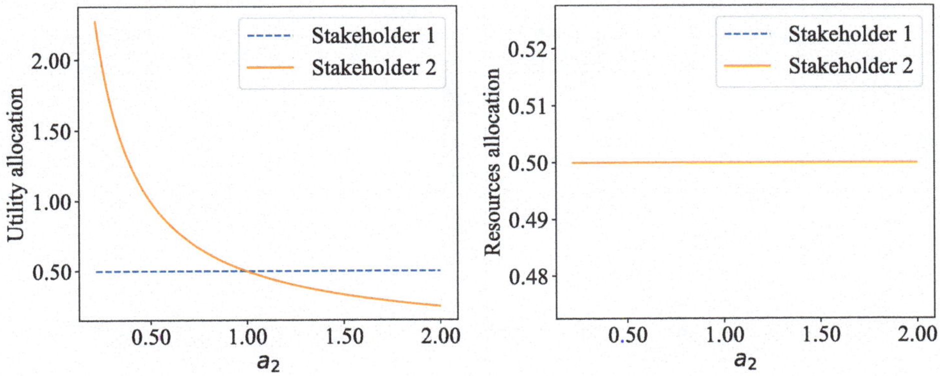

Fig. A.4 Kalai–Smorodinsky SWF. Total budget = 1. Stakeholder 1 coefficient $a_1 = 1$

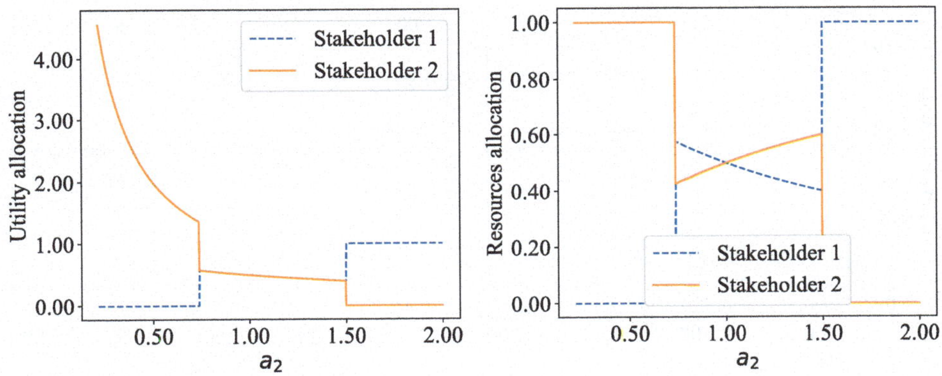

Fig. A.5 Utility threshold SWF. Total budget = 1. Stakeholder 1 coefficient $a_1 = 1$. Delta = 0.2

beta being 0.85. The graph shows that even as Stakeholder 2's costs increase, they continue to receive a substantial allocation of resources. While Stakeholder 1, the more efficient stakeholder, still receives a larger share, alpha fairness ensures that Stakeholder 2 is not drastically disadvantaged despite their higher cost of utility conversion.

Kalai–Smorodinsky SWF (Fig. A.4): The Kalai–Smorodinsky bargaining solution ensures that both stakeholders receive fair, proportional shares based on their respective abilities (in our case, the upper bounds of their utilities). As Stakeholder 2's cost coefficient (a_2) increases, the resource allocation remains balanced due to our numerical setup of equal upper utility bounds.

Utility Threshold SWF (Fig. A.5): In this graph, a utility threshold is applied to ensure that neither stakeholder's utility falls below a certain level ($\Delta = 0.2$). As a_2 increases, Stakeholder 2 becomes more expensive to support, but the threshold ensures that they continue to receive enough resources to maintain a minimum utility. Stakeholder 1 still receives a greater

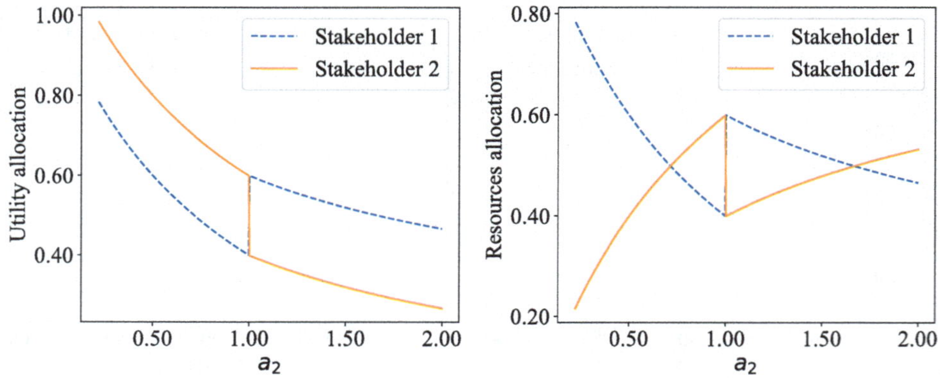

Fig. A.6 Equity threshold SWF. Total budget = 1. Person 1 coefficient $a_1 = 1$. Delta bar = 0.2

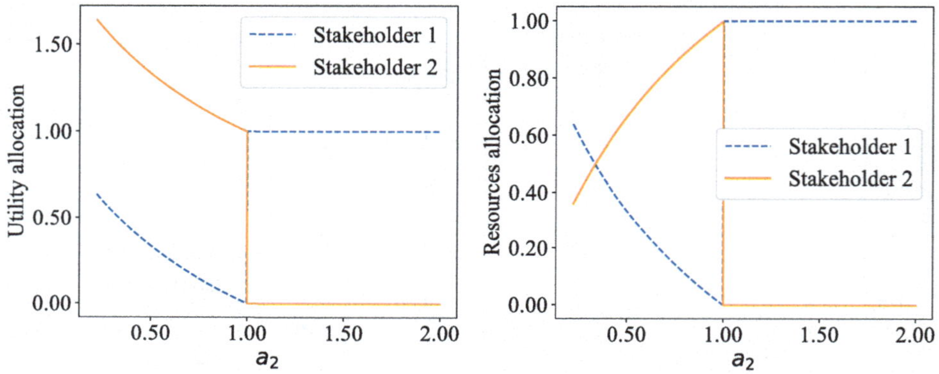

Fig. A.7 Inequality bounds SWF. Total budget = 1. Stakeholder 1 coefficient $a_1 = 1$. Maximum range $R = 1$

share due to their efficiency, but Stakeholder 2 is protected from extreme disadvantages up to a certain level. This SWF switches between utilitarian and maximin.

Equity Threshold SWF (Fig. A.6): Similar to the utility threshold, but focusing on equity, this graph shows that both stakeholders receive resource allocations that keep their utilities relatively close (no more than Δ), despite the increasing cost of converting resources for Stakeholder 2. As a_2 increases, Stakeholder 1, the more efficient stakeholder, receives more resources, but the equity threshold prevents the gap from being more than $\Delta = 0.2$.

Inequality Bounds SWF (Fig. A.7): This graph places an upper bound on the inequality between the two stakeholders. As Stakeholder 2's cost coefficient a_2 increases, Stakeholder 1 begins to receive more resources due to their greater efficiency. However, the SWF ensures that the inequality in utility between the two stakeholders does not exceed a certain level ($R = 1$). This allows for flexibility in allocation while preventing extreme inequality, even as Stakeholder 2 becomes more costly to support.

The manufacturer's authorised representative in the EU is Springer Nature Customer Service Centre GmbH, Europaplatz 3, 69115 Heidelberg, Germany. If you have any concerns regarding our products, please contact ProductSafety@springernature.com

Printed and bound by CPI Group (UK) Ltd, Croydon, CR0 4YY
05/10/2026
02228156-0008